Video over Cognitive Radio Networks

Shiwen Mao

Video over Cognitive Radio Networks

When Quality of Service Meets Spectrum

Springer

Shiwen Mao
Auburn University
Auburn, NY
USA

ISBN 978-1-4614-4956-0 ISBN 978-1-4614-4957-7 (eBook)
DOI 10.1007/978-1-4614-4957-7
Springer New York Heidelberg Dordrecht London

Library of Congress Control Number: 2013956327

Printed on acid-free paper

Springer is part of Springer Science+Business Media (www.springer.com)

To my wife Yihan and son Eric,
for their love and support

–S. Mao

Acknowledgments

I am grateful to Mr. Brett Kurzman and Ms. Rebecca Hytowitz of Springer US. This monograph would be impossible without their patience and continued support during the entire process.

I also thank Dr. Donglin Hu who was involved in this research when he was pursuing his Ph.D. at Auburn University. Special thanks also go to Drs. Thomas Y. Hou and Jeffrey H. Reed for fruitful collaborations during this work.

Shiwen Mao's research is supported in part by the US National Science Foundation (NSF) under Grants CNS-0953513, CNS-1247955, CNS-1320664, and DUE1044021, and through the NSF Broadband Wireless Access & Applications Center (BWAC) Site at Auburn University (NSF Grant IIP-1266036). Any opinions, findings, and conclusions or recommendations expressed in this material are those of the author and do not necessarily reflect the views of the foundation.

Contents

Figures

Table

Chapter 1
Introduction

1.1 Background and Motivation

Due to significant advances in wireless access technologies and the proliferation of wireless devices and applications, there is a fundamental change in wireless network traffic. As predicted by a Cisco study, wireless data is expected to grow to 6.3 Exabytes per month by 2015, a 26-fold increase over 2010. In particular, mobile video is predicted to grow at a compound annual growth rate (CAGR) of 90 % from 2011 to 2016, and 66 % of the increase in future wireless data traffic will be video related [5]. Such dramatic increase in wireless video traffic is driven by the proliferation of mobile PCs, smartphones, tablets, etc., with 300–400 million new mobile phone users adopting mobile services around the world and 120,000 new base stations (BS) deployed every year to meet the compelling need for ubiquitous access of mobile multimedia data.

Such fundamental changes in wireless data volume and composition bring about great challenges for the design and operation of wireless access networks. The capacity of existing and future wireless networks will be greatly stressed. Although allocating more spectrum may help, we are facing the problem of spectrum depletion since it is not a regenerable resource. Improving spectrum efficiency thus becomes ultimately important. In addition, Quality of Service (QoS) /Quality of Experience (QoE) provisioning in wireless networks also becomes a very important problem in order to enable high-quality video services in legacy and emerging wireless networks. Since most of the increase in wireless video will be concentrated in the hot-spot areas, interference becomes the major limiting factor of network capacity and QoS provisioning. Effective interference exploitation and mitigation technologies are needed to achieve more efficient use of the spectrum and power resources.

Among various potential techniques, we consider Cognitive *radios* (CR) as an effective solution to meeting the critical demand in wireless network capacity and enabling wireless video services. Although considerable understandings have been gained on various aspects of CR, the problem of guaranteeing application performance has not been the focus of major CR research. To this end, we find spectrum-intensive and rate-adaptive video, as a reference application, makes excellent use

S. Mao, *Video over Cognitive Radio Networks*, DOI: 10.1007/978-1-4614-4957-7_1,

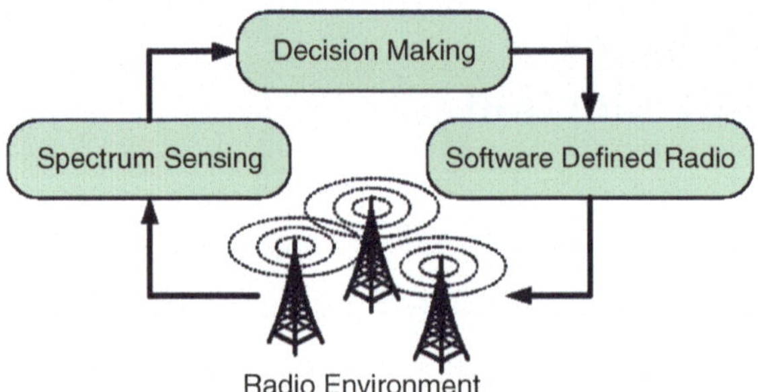

Fig. 1.1 The cognitive-loop

of the enhanced spectrum efficiency in CR networks. Unlike data, where each bit should be delivered, video is loss tolerant and rate adaptive. They are highly suited for CR networks, where the available bandwidth heavily depends on primary user behavior.

1.2 Cognitive Radio in a Nutshell

A CR is a frequency-agile wireless communication device with intelligent control and a monitoring interface that enables efficient and flexible spectrum access [6, 7]. It capitalizes on advances in signal processing and radio technology (i.e., software defined radio (SDR)) [8], as well as in spectrum regulation policy [9, 10]. A CR node can sense the radio environment to detect unused frequency bands (or, *white spaces*) and other state information. A cognitive engine makes intelligent decisions about adapting the radio operation and tuning its parameters. Its frequency-agile radio module is capable of reconfiguring RF and can be programmed to tune to a wide spectrum range and operate on any frequency bands in the range.

The *cognitive loop* of a CR is illustrated in Fig. 1.1, where spectrum sensing, decision making, and spectrum access are the three key components. A truly reconfigurable RF front-end and CR transceiver is indispensable, with flat gain and noise figure and sufficient input return loss over the entire spectrum. The ability to measure, sense, and learn the parameters related to the radio channel characteristics is also essential for a radio to be cognitive. Model uncertainties can significantly affect the performance of sensing schemes, while the sensing accuracy can significantly influence the performance of cognitive networking protocols. The problem is further compounded in the case of multiple channels, multiple primary users, and multiple secondary users. Based on sensing results, the decision-making unit adapts to the radio environment by tuning the SDR and allocating network resources. The new dimension of dynamics on channel availability, sensing, and access makes it a challenging task to develop CR networking protocols with desirable performance. The manifold design trade-offs, multifarious dynamics, and limited resources necessitate a cross-layer optimization approach for "squeezing" the most out of this new paradigm of wireless networking.

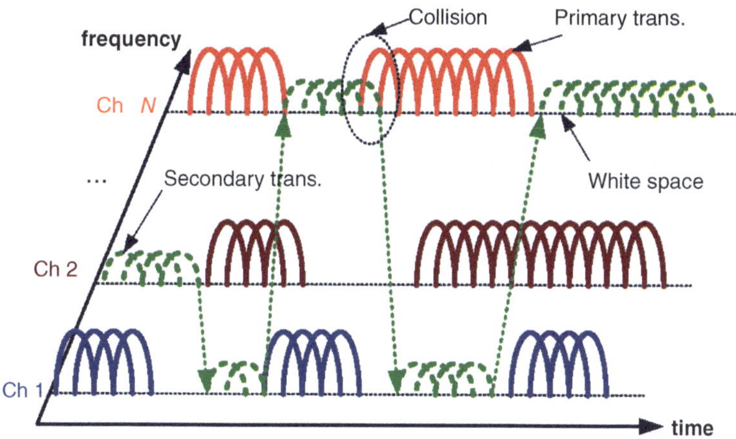

Fig. 1.2 Dynamic spectrum access illustrated

Since Dr. Joseph Mitola III coined the term in the late 1990s [7], CR has attracted tremendous interest from the industry, government, and academia alike. The idea of dynamic spectrum access was largely motivated by FCC's recent finding of serious spectrum underutilization [11], which is in drastic contrast to the common belief about spectrum shortage [12]. Such vast discrepancy calls for an overhaul of the spectrum regulation policy

In some sense, the CR idea is reminiscent of the historic move from *circuit switching* to *packet switching* decades ago, which makes the Internet a reality. A CR network is quite different from traditional networks since it involves users from *multiple* heterogeneous networks sharing the same frequency band(s), with strict priority given to licensed users (or, *primary users*). The CR concept represents a significant paradigm change from exclusive use of spectrum by licensed users, to dynamic spectrum access for unlicensed users (or, *secondary users*). The fundamental objective is to maximize spectrum usage of secondary users in a nonintrusive manner, while protecting primary users from harmful interference. Such a dynamic spectrum access paradigm is illustrated in Fig. 1.2. CR will drastically change the way how wireless systems are designed, and drive the next generation radio devices and standards to enable a broad range of new applications.

As expected, the high potential of CR has stimulated a flurry of exciting activities in engineering, economics, and regulatory communities in searching for better spectrum management policies and sharing techniques. The major technical challenges lie in (i) the tightly coupled networks sharing the same spectrum band, (ii) the tension between primary user protection and secondary user spectrum utilization, and (iii) the unavoidable errors in spectrum sensing and access. Nevertheless, considerable progress has been made and better understandings gained on many aspects of CR, including spectrum white space modeling, spectrum sensing, and spectrum access [12, 13].

1.3 Video over Cognitive Radio Networks

At this juncture, a critical question to ask is "what are we going to do with the extra bandwidth harvested by CR?" Once basic understandings are gained, answers to this question will have profound impact on the future CR research. Historically, CR has evolved from the PHY (e.g., reconfiguring radios on the fly) to higher layers (e.g., network-wide situation awareness). Although we are approaching the Shannon Limit in PHY, there is still huge space for improvement in higher layers rich in open research problems [14], especially in QoS provisioning for multimedia applications.

However, since the need for efficient dynamic spectrum access has driven most of the CR research so far, application performance guarantee has not been the primary concern in mainstream CR research. There are doubts on what applications can be offered. Can such highly dynamic networks carry real-time traffic? Can existing protocols provide satisfactory performance in CR networks? We believe there is compelling need to fully capitalize the CR potential for rich multimedia services in emerging CR networks. Built upon prior research on wireless video, wireless networks and, specifically, CR networks, it is time to investigate the challenging problem of enabling high-quality video service in emerging CR networks. Video is a perfect application for CR, because (i) it is spectrum-intensive (an excellent match for the enhanced spectrum efficiency in CR networks), (ii) it has the most stringent QoS requirements (with certain relaxation, outcomes from this research can be adapted for other types of media, say, VoIP, and applications), and (iii) it is one of the most important applications for the future success and widespread deployment of CR networks. Much research is needed to clear the doubts on the feasibility, create the theoretic foundation, and develop the enabling technologies for video over CR networks.

Although highly rewarding, enabling video in emerging CR networks brings about a whole level of technical challenges. CR introduces a new dimension of dynamics on channel availability, sensing, and access, making it a tremendous task to deliver video data with stringent QoS requirements. To this end, a cross-layer optimization and control approach, complemented with distributed algorithm design, seems to be highly promising. CR video networks provide a perfect setting for the cross-layer optimization and control approach. The close interactions among all the layers (from PHY to application) necessitate cross-layer design, which usually results in highly complex optimization problems. The rapidly varying network environment entails adaptive and robust control. The manifold design trade-offs require judicious selection of operating points and parameters. The inevitable errors in, say, spectrum sensing and sharing, call for algorithms that can provide superior performance and stability in the presence of uncertainties.

1.4 Outline

In this monograph, we investigate the problem of *cross-layer design and optimization of CR video networks*. The objective of this research is two-fold: (i) to develop theoretical bounds and performance limits for future CR video networks with an optimization approach, and (ii) to develop effective algorithms to support video in emerging CR networks. Special consideration will be given to explore the trade-off between maximizing secondary user video quality and primary user protection, as well as quantifying and bounding the effect of sensing errors. We aim to develop a holistic analytical framework that allows the unification of the multidimensional cross-layer design space into a tractable, understandable research problem, complemented with distributed algorithm development for practical CR video systems.

In particular, we adopt scalable video coding, such as fine-grained scalability (FGS) and medium grain scalable (MGS), to encode video streams. We tackle the problems of video over various CR networking paradigms, such as infrastructure-based CR networks, cooperative relay-based CR networks, femtocell-based CR networks, and multi-hop infrastructureless CR networks. We formulate cross-layer optimization problems that incorporate various system parameters and control knobs, and develop effective solution algorithms with proved optimal performance of tight performance bounds.

The remainder of this monograph is organized as follows. In Chap. 2, we examine the problem of video over infrastructure-based CR networks. We consider cross-layer design factors such as scalable video coding, spectrum sensing, opportunistic spectrum access, primary user protection, scheduling, error control, and modulation. We propose efficient optimization and scheduling algorithms for highly competitive solutions, and prove the complexity and optimality bound of the proposed greedy algorithm.

In Chap. 3, we investigate cooperative relay in CR networks using video as a reference application. That is, relays are incorporated to enhance the capacity of the CR cellular network. We consider a base station (BS) and multiple relay nodes (RN) that collaboratively stream multiple videos to CR users within the network. Zero-forcing, an interference alignment technique where the BS and RNs simultaneously transmit encoded mixed signals to all CR users with suitable precoding, such that undesired signals will be canceled and the desired signal can be decoded at each CR user.

In Chap. 4, we investigate the problem of video streaming in femtocell CR networks, where femtocells are deployed to extend coverage, reduce interference, leverage spatial reuse, and enhance network capacity. We formulate a stochastic programming problem for three deployment scenarios. In the case of a single FBS, we apply dual decomposition to develop an optimum-achieving distributed algorithm, which is shown also optimal for the case of multiple non-interfering FBSs. In the case of multiple interfering FBSs, we develop a greedy algorithm that can compute near-optimal solutions, and prove a closed-form lower bound for its performance based on an interference graph model.

In Chap. 5, we examine the challenging problem of video over multi-hop infrastructureless CR networks. We model streaming of concurrent videos as an MINLP problem by adopting the "cut-through switching" amplify-and-forward model. The objective is to maximize the overall received video quality and fairness among the video sessions, while bounding the collision rate with primary users under spectrum sensing errors. We solve the MINLP problem using a centralized sequential fixing algorithm, and derive upper and lower bounds for the objective value. We then apply dual decomposition to develop a distributed algorithm and prove its optimality and convergence conditions.

Finally, we conclude the book in Chap. 6 with a discussion on open research problems and future directions in this interesting and important problem area.

Chapter 2
Video over Cellular CR Networks

In this chapter, we present a study of optimized real-time video multicast in Cognitive radios (CR) networks. Consider an infrastructure-based CR network collocated with N primary networks, as illustrated in Fig. 2.1. Each primary network is allocated with a channel. The availability of each channel evolves over time due to primary user transmissions. We consider multicast application due to its generality and bandwidth efficiency. The CR base station exploits spectrum opportunities in the N channels to multicast videos to G multicast groups. In order to accommodate heterogeneous user channels, fine-grained scalability (FGS) is adopted to encode each video into a base layer and an enhancement layer [15]. With FGS, the enhancement layer can be truncated at any bit location, while the remaining bits are all useful for decoding. Therefore a user can receive a video quality commensurate to its channel condition.

We present a formulation for the CR video multicast problem, taking into account various cross-layer design factors such as scalable video coding, spectrum sensing, dynamic spectrum access, modulation, scheduling, error control, and primary user protection. The objective is three-fold: (i) to optimize the overall received video quality, (ii) to achieve proportional fairness among multicast users, and (iii) to protect primary users from harmful collisions. Unlike prior work on wireless video [15, 16], the challenge stems from dynamic channel availability processes, tightly coupled design choices, and the need to predict future channel status under the presence of sensing errors for partitioning, modulation and scheduling of real-time video data.

The formulated problem is shown to be a mixed integer nonlinear programming (MINLP) problem, which usually has high complexity to solve. However, the partitioning and scheduling of video packets to the multiple channels should be performed in real-time. This is because the availability of the channels is determined by primary user transmissions, which may occur in any time slot, and DSA should be unobtrusive to primary users. Any change in channel status will affect both video data partitioning and packet scheduling. The need for real-time execution calls for low-complexity algorithms. For performance, we aim to design algorithms with proven optimality bounds for video applications.

S. Mao, *Video over Cognitive Radio Networks*, DOI: 10.1007/978-1-4614-4957-7_2,

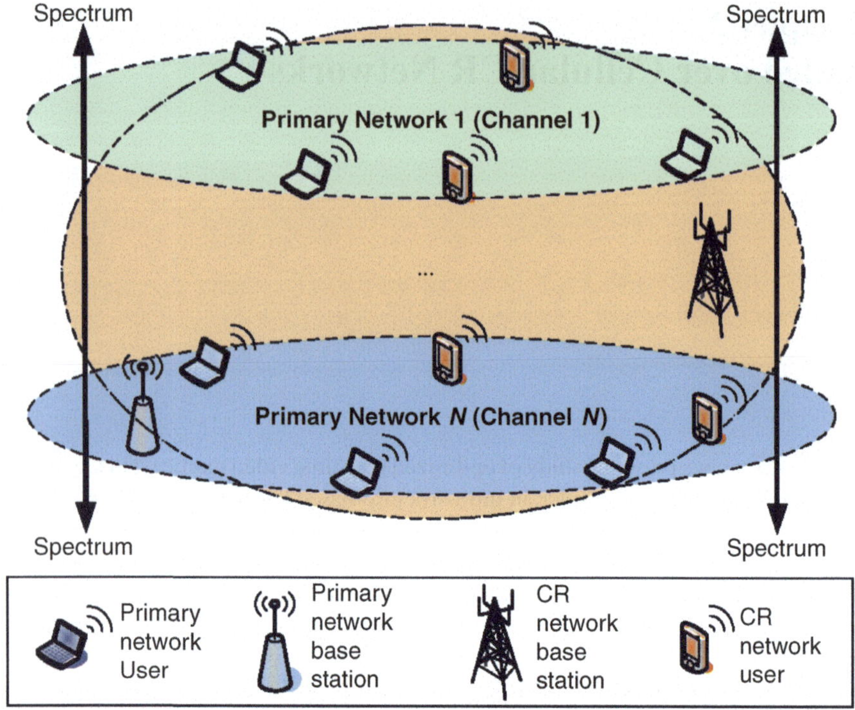

Fig. 2.1 Illustration of the network architecture considered in this chapter: an infrastructure-based CR network collocated with N primary networks ([1], ©2010 IEEE)

We then develop a two-step approach to solve the formulated MINLP problem. For each group of pictures (GoP), we first determine the optimal partition (and thus rates and modulation-coding (MC) schemes) of FGS video data. We present two computationally efficient algorithms for this purpose: (i) a *sequential fixing* (SF) algorithm based on a linear relaxation of the MINLP problem [17]; and (ii) a *greedy algorithm*, termed GRD1, which exploits the inherent priority structure of FGS video and the order of user channels according to their qualities. We show that GRD1 can guarantee a solution that is within a factor (i.e., approximation ratio) of $(1 - e^{-1/2})$ of the global optimal solution, while with a polynomial complexity suitable for execution of each GoP. The computed solution is further adjusted in each time slot based on more recent channel sensing results and feedback, using a *refined greedy algorithm*, termed GRD2, with a further reduced complexity suitable for execution in each time slot. During each time slot, we use a *tile scheduling algorithm*, termed TSA, to assign video packets (the amount of which is determined by GRD1 and GRD2) to available channels, while each channel is accessed with a probability derived from spectrum sensing results. TSA has a low complexity of $O(N \log N)$ and is optimal in terms of maximizing the total utility of the users.

Simulation results are presented to provide a comparison study with alternative schemes as well as demonstrating the impact of several key design parameters on the overall system performance. It is observed that excellent performance can be achieved by the proposed algorithms, with considerable improvement over an alternative equal allocation scheme. The opportunistic spectrum accessapproach makes FGS video robust to sensing errors.

The rest of this chapter is organized as follows. In Sect. 2.1, we present the CR video multicast framework. The proposed algorithms are presented in Sect. 2.2. We show our simulation study in Sect. 2.3 and discuss the related work in Sect. 2.4. Section 2.5 concludes the chapter.

2.1 CR Video Multicast Framework

2.1.1 Network Model

Primary Networks

Consider a spectrum band consisting of N channels, each evolving over time independently. As illustrated in Fig. 2.1, we assume that the N channels are allocated to N primary networks. For ease of presentation, we assume that primary users access the channels following a synchronous slot structure [12, 18]. The status of each of the N channels evolves following a two-state discrete-time Markov process [12, 19]. The network status in slot t is

$$\vec{S}(t) = [S_1(t), S_2(t), \ldots, S_N(t)],$$

where $S_n(t)$ denotes the status of channel n with idle ($S_n(t) = 0$) and busy ($S_n(t) = 1$) states. Let λ_n be the transition probability of remaining in the idle state, and μ_n the transition probability from the busy state to the idle state for channel n.

CR Cellular Network

Consider a CR network collocated with the N primary networks, within which a base station multicasts G real-time videos to G multicast groups, each having N_g users, $g = 1, 2, \ldots, G$. The base station seeks spectrum opportunities in the N channels. In each time slot t, the base station chooses a set of channels $\mathcal{A}_1(t)$ to sense and a set of channels $\mathcal{A}_2(t)$ to access. The base station has $|\mathcal{A}_1(t)|$ transceivers such that it can sense $|\mathcal{A}_1(t)|$ channels simultaneously. A time slot and channel combination, termed a *tile*, is the minimum unit for resource allocation.

The same time-slot structure is adopted as in [12, 18], which is illustrated in Fig. 2.2. At the beginning of each time slot, the base station senses channels in

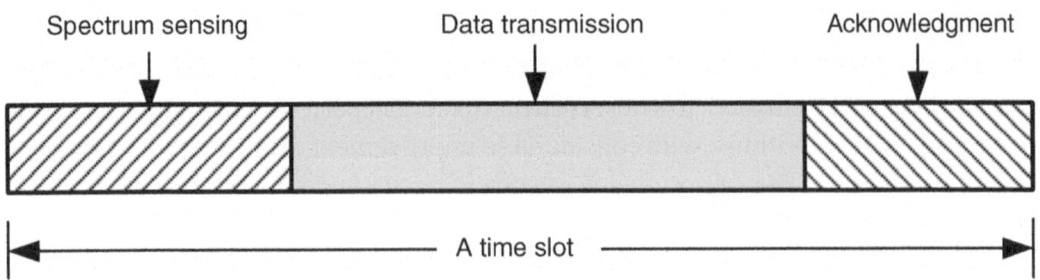

Fig. 2.2 The structure of a time slot ([1], ©2010 IEEE)

$\mathcal{A}_1(t)$ and then chooses a set of available channels for opportunistic transmissions based on sensing results. After a successful transmission, the base station will receive an ACK from the user with the highest SNR in the target multicast group.[1]

Without loss of generality, we assume that each CR network user can access all the N channels. We adopt OFDM as multicast technology at the PHY layer. In OFDM, the spectrum is divided into narrowband channels and the signals are modulated on the channels in the frequency domain. An OFDM frame can be transmitted through one antenna and received through one antenna, since the symbols can be modulated to multiple channels by inverse fast Fourier transform (IFFT). Since it is always desirable to have low hardware requirements, we assume each CR network user has one antenna. The adaptability of OFDM makes it highly suitable for CR networks [12, 21, 22].

2.1.2 Spectrum Sensing

Although precise and timely channel state information is desirable for spectrum access and primary user protection, continuous full-spectrum sensing is both energy inefficient and hardware demanding. We assume $|\mathcal{A}_1(t)|$ channels are sensed in each time slot, while sensing is carried out on every $W = N/|\mathcal{A}_1(t)|$ channels [18]. The indices of the channels to be sensed are

$$n = (hW + t) \bmod (N + 1), \quad h = 0, 1, \ldots, |\mathcal{A}_1(t)| - 1, \tag{2.1}$$

where t is the discrete time index. The base station senses the channels in an increasing order. At the beginning of a slot, the base station chooses $|\mathcal{A}_1(t)|$ channels to sense following (2.1). It then predicts channel status based on sensing results and channel history.

[1] Although ACKs are not adopted in many multicast applications, it has been shown that the *feedback implosion* problem can be effectively solved using properly designed timers [20] (e.g., reversely proportional to channel SINR). The ACKs are important for predicting future channel status (see Sect. 2.1.2).

During the sensing process, two kinds of detection errors may occur. When there is a *false alarm*, a spectrum opportunity will be wasted. When there is a *miss detection*, the base station may make a transmission on a busy channel and thus cause collision with primary users. Let ϵ_n and δ_n denote the probabilities of false alarm and miss detection on channel n, respectively. The spectrum sensing performance can be represented by the receiver operation characteristic (ROC) curve, which gives $1-\delta_n$ as a function of ϵ_n [12, 23]. Let $\vec{R}(t)$ be the sensing outcome in slot t, as

$$\vec{R}(t) = [R_1(t), R_2(t), \ldots, R_N(t)].$$

If channel n is not sensed in time slot t, we have $R_n(t) = -1$. If channel n is sensed in slot t, it has value of either $R_n(t) = 1$ (i.e., sensed busy) or $R_n(t) = 0$ (i.e., sensed idle). The sensing error probabilities for a sensed channel n are $P(R_n(t) = 1|S_n(t) = 0) = \epsilon_n$ and $P(R_n(t) = 0|S_n(t) = 1) = \delta_n$.

Let $\vec{a}(t) = [a_1(t), a_2(t), \ldots, a_N(t)]$ be the *belief vector*, where each element is the conditional probability $a_n(t) = P(S_n(t) = 0|\theta_n(t))$ and $\theta_n(t)$ is defined as the channel n history up to the end of the sensing stage of time slot t. Furthermore, let $\check{a}_n(t) = P(S_n(t) = 0|\check{\theta}_n(t))$ and $\check{\theta}_n(t)$ is the channel n history up to the end of the ACK stage of time slot t. $a_n(t)$ and $\check{a}_n(t)$ are conditional probabilities based on past sensing results and transmission results on channel n, respectively. We have for time slot t

$$\check{a}_n(t) = \begin{cases} 1, & \text{transmission in slot } t \textit{ and } \text{ACK received} \\ 0, & \text{transmission in slot } t \textit{ and } \text{no ACK received} \\ a_n(t), & \text{no transmission in slot } t. \end{cases} \quad (2.2)$$

The belief vector $\vec{a}(t)$ can be estimated by considering the following three cases. Case I, channel n is not sensed in time slot t. We need to estimate $a_n(t)$ from the channel n history in the previous time slot. We have

$$\begin{aligned} a_n(t) &= P(S_n(t) = 0|R_n(t) = -1, \theta_n) \\ &= \lambda_n \check{a}_n(t-1) + \mu_n[1 - \check{a}_n(t-1)] \\ &= \pi_n(t). \end{aligned} \quad (2.3)$$

Case II, channel n is sensed in time slot t and the sensing result is $R_n(t) = 0$. The availability of channel n in time slot t is then conditioned on both the channel history and the sensing result. We have

$$\begin{aligned} a_n(t) &= P(S_n(t) = 0|R_n(t) = 0, \theta_n) \\ &= \frac{P(S_n(t) = 0, R_n(t) = 0|\theta_n)}{\sum_{X_n \in \{0,1\}} P(S_n(t) = X_n, R_n(t) = 0|\theta_n)} \\ &= \frac{P(R_n(t) = 0|S_n(t) = 0) P(S_n(t) = 0|\theta_n)}{\sum_{X_n} P(R_n(t) = 0|S_n(t) = X_n) P(S_n(t) = X_n|\theta_n)} \end{aligned}$$

$$= \frac{\pi_n(t)(1-\epsilon_n)}{\pi_n(t)(1-\epsilon_n) + [1-\pi_n(t)]\delta_n}. \quad (2.4)$$

The third step in (2.4) is due to the memoryless property of the channel process.

Case III, channel n is sensed in time slot t and the sensing result is $R_n(t) = 1$. Similarly, we can derive the expression for $a_n(t)$ for this case as

$$\begin{aligned} a_n(t) &= P(S_n(t) = 0 | R_n(t) = 1, \theta_n) \\ &= \frac{\pi_n(t)\epsilon_n}{\pi_n(t)\epsilon_n + [1-\pi_n(t)](1-\delta_n)}. \end{aligned} \quad (2.5)$$

2.1.3 Opportunistic Spectrum Access

At the beginning of each time slot t, the CR base station (BS) senses the M channels and computes $a_n(t)$ for each channel n. Based on spectrum sensing results, the BS determines which channels to access for video streaming. We adopt an opportunistic spectrum access approach, aiming to exploit unused spectrum while probabilistically bounding the interference to primary users.

Let $\gamma_n \in (0, 1)$ be the *maximum allowed collision probability* with primary users on channel n, and $p_n^{\text{tr}}(t)$ the *transmission probability* on channel n for the base station in time slot t. The probability of collision caused by the base station should be kept below γ_n, i.e., $p_n^{\text{tr}}(t)\,[1 - a_n(t)] \leq \gamma_n$. In addition to primary user protection, another important objective is to exploit unused spectrum as much as possible. The transmission probability can be determined by jointly considering both objectives, as

$$p_n^{\text{tr}}(t) = \begin{cases} \min\left\{1, \frac{\gamma_n}{1-a_n(t)}\right\}, & \text{if } 0 \leq a_n(t) < 1 \\ 1, & \text{if } a_n(t) = 1. \end{cases} \quad (2.6)$$

If $p_n^{\text{tr}}(t) = 1$, channel n will be accessed deterministically. If $p_n^{\text{tr}}(t) = \gamma_n / [1 - a_n(t)] < 1$, channel n will be accessed opportunistically with probability $p_n^{\text{tr}}(t)$.

2.1.4 Modulation-Coding Schemes

At the PHY layer, we consider various modulation and channel coding combination schemes. Without loss of generality, we assume several choices of modulation schemes, such as QPSK, 16-QAM, and 64-QAM, combined with several choices of forward error correction (FEC) schemes, e.g., with rates 1/2, 2/3, and 3/4. We consider M unique combinations of modulation and FEC schemes, termed *Modulation-Coding* (MC) schemes, in this chapter.

Under the same channel condition, different MC schemes will achieve different data rates and symbol error rates. Adaptive modulation and channel coding allow us to exploit user channel variations to maximize video data rate under a given residual bit error rate constraint. When a user has a good channel, it should adopt an MC scheme that can support a higher data rate. Conversely, it should adopt a low-rate MC scheme when the channel condition is poor. Let $\{MC_m\}_{m=1,\ldots,M}$ be the list of available MC schemes indexed according to their data rates in the increasing order. We assume slow fading channels with coherence time larger than a time slot. Each CR user measures its own channel and feedbacks measurements to the base station when its channel quality changes. At the beginning of a time slot, the base station is able to collect the number $n_{g,m}$ of users in each multicast group g that can successfully decode MC_m signals for $m = 1, 2, \ldots, M$.

Since the base layer carries the most important data, the most reliable MC scheme $MC_{b(g)}$ should be used, where $b(g) = \max_i\{i : n_{g,i} = N_g\}$, for all g. Without loss of generality, we assume that the base layer is always transmitted using MC_1. If a user's channel is so poor that it cannot decode the MC_1 signal, we consider it disconnected from the CR network. We further divide the enhancement layer into M sub-layers, where sub-layer m has rate $R^e_{g,m}$ and uses MC_m. Assuming that MC_m can carry $b_{g,m}$ bits of video g in one tile, we denote the number of tiles for sub-layer m of video g as $l_{g,m} \geq 0$. We have

$$R^e_g = \sum_{m=1}^{M} R^e_{g,m} = \sum_{m=1}^{M} b_{g,m} l_{g,m}. \tag{2.7}$$

2.1.5 Proportional Fair Allocation

Since video quality is considered in this chapter, we define the utility for user i in group g as

$$U_{g,i} = \log Q_{g,i} = \log\left(Q^b_g + \beta_g R^e_g(i)\right),$$

where $R^e_g(i)$ is the received enhancement layer rate of user i in group g.

The total utility for group g is

$$U_g = \sum_{i=1}^{N_g} U_{g,i}$$

Intuitively, a lower layer should use a lower (i.e., more reliable) MC scheme. This is because if a lower layer is lost, a higher layer cannot be used at the decoder even if it is correctly received. Considering the user classification based on their MC schemes, we can rewrite U_g as follows [24]:

$$U_g = \sum_{k=1}^{M} (n_{g,k} - n_{g,k+1}) \log\left(Q_g^b + \beta_g \sum_{m=1}^{k} R_{g,m}^e \right), \tag{2.8}$$

where $n_{g,M+1} = 0$. The utility function of the entire CR video multicast system is

$$U = \sum_{g=1}^{G} U_g. \tag{2.9}$$

Maximizing U will achieve *proportional fairness* among the video sessions [25]

2.2 Optimized Video Multicast in CR Networks

2.2.1 Outline of the Proposed Approach

As discussed, the CR video multicast problem is highly challenging since a lot of design choices are tightly coupled. First, as users see different channels, such heterogeneity should be accommodated so that a user can receive a video quality commensurate to its channel quality. Second, we need to determine the video rates before transmission, which, however, depend on future channel evolution and choice of MC schemes. Third, the trade-off between primary user protection and spectrum utilization should guide the scheduling of video packets to channels. Finally, all the optimization decisions should be made in real-time. Low-complexity, but efficient algorithms are needed, while theoretical optimality bounds would be highly appealing.

To address heterogeneous user channels, we adopt FGS to produce a base layer with rate R_g^b and an enhancement layer with rate $\bar{R}_g^e$. Without loss of generality, we assume R_g^b is prescribed for an acceptable video quality, while $\bar{R}_g^e$ is set to a large value that is allowed by the codec. During transmission, we determine the *effective rate* for each enhancement layer $R_g^e \le \bar{R}_g^e$ depending on channel availability, sensing, and MC schemes.[2] The optimal partition of the enhancement layer should be determined such that each sub-layer uses a different MC scheme.

We determine the optimal partition of enhancement layers, the choices of MC schemes, and video packet scheduling as follows. First, we solve the optimal partition problem for every GoP based on an estimated (i.e., average) number of available tiles T_e in the next GoP window that can be used for the enhancement layer, using algorithm GRD1 with complexity $O(\text{MGT}_e)$. The tile allocations are then dynamically adjusted in each time slot according to more recent (and thus more accurate) channel status using algorithm GRD2, with complexity $O(\text{MGK})$, where $K \ll T_e$. Second, during each time slot, video packets are scheduled to the available

[2] The proposed approach can also be used for streaming stored FGS video.

channels such that the overall system utility is maximized. The TSA algorithm has complexity $O(N \log N)$. Both GRD2 and TSA have low complexity and are suitable for execution in each time slot.

In real-time video, overdue packets generally do not contribute to improving the received quality. We assume that the data from a GoP should be be delivered in the next GoP window consisting of T_{GoP} time slots.[3] Since the base layer is essential for decoding a video, we assume that the base layers of all the videos are coded using MC_1. For the M sub-layers of the enhancement layer, a more important sub-layer will be coded using a more reliable (i.e., lower rate) MC scheme. At the beginning of each GoP window, all the base layers are transmitted using the available tiles. *Retransmissions* will be scheduled if no ACK is received for a base layer packet. After the base layers are transmitted, we allocate the remaining available tiles in the GoP window for the enhancement layer. The same rule applies to the enhancement sub-layers, such that a higher sub-layer will be transmitted if and only if all the lower sub-layers are acknowledged. This is due to the decoding dependency of layered video.

In each time slot t, the base station opportunistically access every channel n with probability $p_n^{\text{tr}}(t)$ given in (2.6). Specifically, for each channel n, the base station generates a random number $x_n(t)$, which is independent of the channel history $\theta_n(t)$ and uniformly distributed in [0,1]. If $x_n(t) \leq p_n^{\text{tr}}(t)$, the most important packet among those not ACKed in the previous GoP will be transmitted on channel n. If an ACK is received for this packet at the end of time slot t, this packet is successfully received by at least one of the users and will be removed from the transmission buffer. Otherwise, there is a collision with primary user and this packet will remain in the transmission buffer and will be retransmitted.

In the following, we describe in detail the three algorithms.

2.2.2 Enhancement Layer Partitioning and Tile Allocation

As a first step, we need to determine the effective rate for each enhancement layer $R_g^e \leq \bar{R}_g^e$. We also need to determine the optimal partition of each enhancement layer. Clearly, the solutions will be highly dependent on the channel availability processes and sensing results.

Recall that the base layers are transmitted using MC_1 first in each GoP window. The *remaining* available tiles can then be allocated to the enhancement layers. We assume that the number of tiles used for the enhancement layers in a GoP window, T_e, is known at the beginning of the GoP window. For example, we can estimate T_e by computing the total average "idle" intervals of all the N channels based on the channel model, decreased by the number of tiles used for the base layers

[3] The proposed approach also works for the more general delay requirements that are multiple GoP windows.

(i.e., $R_g^b/b_{g,1}$). We then split the enhancement layer of each video g into M sub-layers, each occupying $l_{g,m}$ tiles when coded with MC_m, $m = 1, 2, \ldots, M$.

Letting $\vec{l} = [l_{1,1}, l_{1,2}, \ldots, l_{1,M}, l_{2,1}, \ldots l_{G,M}]$ denote the *tile allocation vector*, we formulate an optimization problem OPT-Part as follows:

$$\text{maximize:} \quad U(\vec{l}) =$$

$$\sum_{g=1}^{G} \sum_{k=1}^{M} (n_{g,k} - n_{g,k+1}) \times \log \left[Q_g^b + \beta_g \sum_{m=1}^{k} b_{g,m} l_{g,m} \right] \tag{2.10}$$

subject to:

$$\sum_{g=1}^{G} \sum_{m=1}^{M} l_{g,m} \leq T_e \tag{2.11}$$

$$\sum_{m=1}^{M} b_{g,m} l_{g,m} \leq \bar{R}_g^e, \;\; g \in [1, \ldots, M] \tag{2.12}$$

$$l_{g,m} \geq 0, \;\; m \in [1, \ldots, M], g \in [1, \ldots, G]. \tag{2.13}$$

OPT-Part is solved at the beginning of each GoP window to determine the optimal partition of the enhancement layer. The objective is to maximize the overall system utility by choosing optimal values for the $l_{g,m}$'s. We can derive the effective video rates as $R_g^e = \sum_{m=1}^{M} b_{g,m} l_{g,m}$. The formulated problem is a MINLP problem, which is NP-hard [24]. In the following, we present two algorithms for computing near-optimal solutions to problem OPT-Part: (i) a *sequential fixing* (SF) algorithm based on a linear relaxation of (2.10), and (ii) a *greedy algorithm* GRD1 with proven optimality gap.

A Sequential Fixing Algorithm

With this algorithm, the original MINLP is first linearized to obtain a linear programming (LP) relaxation. Then we iteratively solve the LP, while fixing one integer variable in every iteration [17, 26]. We use the *Reformulation-Linearization Technique* (RLT) to obtain the LP relaxation [27]. RLT is a technique that can be used to produce LP relaxations for a nonlinear, nonconvex polynomial programming problem. This relaxation will provide a tight upper bound for a maximization problem. Specifically, we linearize the logarithm function in (2.10) over some suitable, tightly bounded interval using a polyhedral outer approximation comprising a convex envelope in concert with several tangential supports. We further relax the integer constraints, i.e., allowing the $l_{g,m}$'s to take fractional values. Then we obtain an upper bounding LP relaxation that can be solved in polynomial time. Due to lack of space, we refer interested readers to [27] for a detailed description of the technique.

Algorithm 1: Sequential Fixing (SF) Algorithm ([1], ©2010 IEEE)

```
1  Use RLT to linearize the original problem ;
2  Solved the LP relaxation ;
3  Suppose l_{ĝ,m̂} is the integer variable with the minimum, (⌈l_{ĝ,m̂}⌉ − l_{ĝ,m̂}) or (l_{ĝ,m̂} − ⌊l_{ĝ,m̂}⌋)
   value among all l_{g,m} variables that remain to be fixed, round it up or down to the nearest
   integer ;
4  if (all l_{g,m}'s are fixed) then
5  |  Go to Step 10 ;
6  else
7  |  Reformulate a new relaxed LP with the newly fixed l_{g,m} variables ;
8  |  Go to Step 2 ;
9  end
10 Output all fixed l_{g,m} variables and R^e_g = Σ_{m=1}^{M} b_{g,m} l_{g,m} ;
```

We next solve the LP relaxation iteratively. During each iteration, we find the $l_{\hat{g},\hat{m}}$ which has the minimum value for $\left(\lceil l_{\hat{g},\hat{m}} \rceil - l_{\hat{g},\hat{m}}\right)$ or $\left(l_{\hat{g},\hat{m}} - \lfloor l_{\hat{g},\hat{m}} \rfloor\right)$ among all fractional $l_{g,m}$'s, and round it up or down to the nearest integer. We next reformulate and solve a new LP with $l_{\hat{g},\hat{m}}$ fixed. This procedure repeats until all the $l_{g,m}$'s are fixed. The complete SF algorithm is given in Algorithm 1. The complexity of SF depends on the specific LP algorithm (e.g., the *simplex method* with polynomial-time average-case complexity).

A Greedy Algorithm

Although SF can compute a near-optimal solution in polynomial time, it does not provide any guarantee on the optimality of the solution. In the following, we describe a greedy algorithm, termed GRD1, which exploits the inherent priority structure of layered video and MC schemes and has a proven optimality bound.

The complete greedy algorithm is given in Algorithm 2, where $R = \sum_{g=1}^{G} \bar{R}^e_g$ is the total rate of all the enhancement layers and $\vec{e}_i$ is a *unit vector* with "1" at the i-th location and "0" at all other locations. In GRD1, all the $l_{g,m}$'s are initially set to 0. During each iteration, one tile is allocated to the $\hat{m}$-th sub-layer of video $\hat{g}$. In Step 4, $l_{\hat{m},\hat{g}}$ is chosen to be the one that achieves the largest increase in terms of the "normalized" utility (i.e., $[U(\vec{l}+\vec{e}_{g,m}) - U(\vec{l})]/[b_{g,m} + R/T_e])$ if it is assigned with an additional tile. Lines 6, 7, and 8 check if the assigned rate exceeds the maximum rate $\bar{R}^e_g$. GRD1 terminates when either all the available tiles are used or when all the video data are allocated with tiles. In the latter case, all the videos are transmitted at full rates. We have the following Theorem for GRD1.

Theorem 2.1 *The greedy algorithm GRD1 shown in Algorithm 2 has a complexity* $O(\mathrm{MGT}_e)$. *It guarantees a solution that is within a factor of* $(1-e^{-1/2})$ *of the global optimal solution.*

Algorithm 2: Greedy Algorithm (GRD1) ([1], ©2010 IEEE)

1 Initialize $l_{g,m} = 0$ for all g and m ;
2 Initialize $A = \{1, 2, \ldots, G\}$;
3 **while** $\left(\sum_{g=1}^{G} \sum_{m=1}^{M} l_{g,m} \le T_e \textit{ and } A \textit{ is not empty}\right)$ **do**
4   Find $l_{\hat{g},\hat{m}}$ that can be increased by one: $\vec{e}_{\hat{g},\hat{m}} = \arg\max_{g \in A, m \in [1,\ldots,M]} \left\{ \frac{U(\vec{l}+\vec{e}_{g,m})-U(\vec{l})}{b_{g,m}+R/T_e} \right\}$;
5     $\vec{l} = \vec{l} + \vec{e}_{\hat{g},\hat{m}}$;
6   **if** $\left(\sum_m b_{\hat{g},m} l_{\hat{g},m} > \bar{R}_g^e\right)$ **then**
7         $\vec{l} = \vec{l} - \vec{e}_{\hat{g},\hat{m}}$;
8   Delete $\hat{g}$ from A ;
9   **end**
10 **end**

Proof (i) *Complexity*: In Step 4 in Algorithm 2, it takes $O(MG)$ to solve for $\vec{e}_{\hat{g},\hat{m}}$. Since each iteration assigns one tile to sub-layer $\hat{m}$ of group $\hat{g}$, it takes T_e iterations to allocate all the available tiles in a GoP window. Therefore, the overall complexity of GRD1 is $O(\text{MGT}_e)$.

(ii) *Optimality Bound*: This proof is extended from a result first shown in [24] for layered videos. We first show a property of group utility $U_g(\vec{l})$, which will be used in the proof of the optimality gap. For two vectors $\vec{l}_g^1$ and $\vec{l}_g^2$, letting $\Delta = U_g(\vec{l}_g^1) - U_g(\vec{l}_g^2)$, we have

$$
\begin{aligned}
\Delta &= \sum_{k=1}^{M} (n_{g,k} - n_{g,k+1}) \times \log\left(1 + \frac{\sum_{m=1}^{k} \beta_g b_{g,m} (l_{g,m}^1 - l_{g,m}^2)}{Q_g^b + \sum_{m=1}^{k} \beta_g b_{g,m} l_{g,m}^2}\right) \\
&\le \sum_{k=1}^{M} \sum_{m=1}^{k} (l_{g,m}^1 - l_{g,m}^2)^+ (n_{g,k} - n_{g,k+1}) \\
&\quad \times \log\left(1 + \beta_g b_{g,m} \Big/ \left[Q_g^b + \sum_{m=1}^{k} \beta_g b_{g,m} l_{g,m}^2\right]\right) \\
&\le \sum_{k=1}^{M} \sum_{m=1}^{M} (l_{g,m}^1 - l_{g,m}^2)^+ (n_{g,k} - n_{g,k+1}) \\
&\quad \times \log\left(1 + \beta_g b_{g,m} \Big/ \left[Q_g^b + \sum_{m=1}^{k} \beta_g b_{g,m} l_{g,m}^2\right]\right) \\
&= \sum_{m=1}^{M} (l_{g,m}^1 - l_{g,m}^2)^+ \left[U_g(\vec{l}_g^2 + b_{g,m}) - U(\vec{l}_g^2)\right],
\end{aligned}
\tag{2.14}
$$

where $y^+ = \max\{0, y\}$. The first inequality is due to the concavity of logarithm functions.

Next we prove the optimality bound. Let $\vec{l}_t$ be the output of GRD1 after t iterations. Let the utility gap between the optimal solution and the GRD1 solution be $F_t = U(\vec{l}^*) - U(\vec{l}_t)$, and $\vec{e}_{\hat{g},\hat{m}}(t)$ the argument found in Step 4 of GRD1 after t iterations. We have $\vec{l}_t = \vec{l}_{t-1} + \vec{e}_{\hat{g},\hat{m}}(t)$ and

$$
\begin{aligned}
&F_{t-1} \\
&= U(\vec{l}^*) - U(\vec{l}_{t-1}) \\
&\leq \sum_g \sum_m (l^*_{g,m} - l_{g,m})^+ \left[U(\vec{l}_{t-1} + \vec{e}_{g,m}(t)) - U(\vec{l}_{t-1}) \right] \\
&\leq \sum_g \sum_m \left(l^*_{g,m} - l_{g,m} \right)^+ \left[U(\vec{l}_{t-1} + \vec{e}_{\hat{g},\hat{m}}(t)) - U(\vec{l}_{t-1}) \right] \frac{b_{g,m} + R/T_e}{b_{\hat{g},\hat{m}}(t) + R/T_e} \\
&\leq \frac{U(\vec{l}_t) - U(\vec{l}_{t-1})}{b_{\hat{g},\hat{m}}(t) + R/T_e} \sum_g \sum_m \left[l^*_{g,m}(b_{g,m} + R/T_e) \right].
\end{aligned}
$$

The first inequality is due to (2.14) and the second inequality follows Step 4 of GRD1. It follows (2.11) that $\sum_g \sum_m l^*_{g,m} \leq T_e$ and $\sum_g \sum_m b_{g,m} l^*_{g,m} \leq R$. We have $F_{t-1} \leq (F_{t-1} - F_t) \frac{2R}{b_{\hat{g},\hat{m}}(t) + R/T_e}$. Solving for F_t, we have

$$
F_t \leq F_{t-1} \left\{ 1 - \left[b_{\hat{g},\hat{m}}(t) + R/T_e \right] / (2R) \right\}.
$$

Suppose the *WHILE* loop in Algorithm 2 has been executed k times when the solution is obtained.

$$
\begin{aligned}
F_k &\leq F_{k-1} \left\{ 1 - \left[b_{\hat{g},\hat{m}}(k) + R/T_e \right] / (2R) \right\} \\
&\leq F_0 \prod_{t=1}^{k} \left\{ 1 - \left[b_{\hat{g},\hat{m}}(t) + R/T_e \right] / (2R) \right\} \\
&\leq F_0 \left\{ 1 - 1/(2kR) \sum_{t=1}^{k} [b_{\hat{g},\hat{m}}(t) + R/T_e] \right\}^k .
\end{aligned}
$$

The *WHILE* loop exits when one or both the constraints are violated. If the constraint $\sum_g \sum_m l_{g,m} \leq T_e$ is violated, there is no tile that can be used. Therefore $k \geq T_e$ and $\sum_{t=1}^{k} R/T_e \geq R$. If constraint "$A$ is not empty" is violated, all the videos have been allocated sufficient number of tiles and will be transmitted at full rates. We have $\sum_{t=1}^{k} b_{\hat{g},\hat{m}}(t) \geq R$ in this case. It follows that

$$
\begin{aligned}
F_k &\leq F_0 \left\{ 1 - 1/(2kR) \sum_{t=1}^{k} [b_{\hat{g},\hat{m}}(t) + R/T_e] \right\}^k \\
&\leq F_0 \left[1 - 1/(2k) \right]^k \;\leq\; F_0 e^{-1/2}.
\end{aligned}
$$

Since $F_0 = U(\vec{l}^*)$, we have $U(\vec{l}_k) \geq (1 - e^{-1/2})U(\vec{l}^*)$. Therefore, we conclude that the GRD1 solution is bounded by $(1 - e^{-1/2})U(\vec{l}^*)$ and $U(\vec{l}^*)$. □

A Refined Greedy Algorithm

GRD1 computes $l_{g,m}$'s based on an estimate of network status $\vec{S}(t)$ in the next T_{GoP} time slots. Due to channel dynamics, the computed $l_{g,m}$'s may not be exactly accurate, especially when T_{GoP} is large. We next present a refined greedy algorithm, termed GRD2, which adjusts the $l_{g,m}$'s based on more accurate estimation of the channel status.

GRD2 is executed at the beginning of every time slot. It estimates the number of available tiles $T_e(t)$ in the next T_{est} successive time slots, where $1 \leq T_{\text{est}} \leq T_{\text{GoP}}$ is a design parameter depending on the coherence time of the channels. Such estimates are more accurate than that in GRD1 since they are based on recently received ACKs and recent sensing results. Specifically, we estimate $T_e(t)$ using the belief vector $\vec{a}(t)$ in time slot t. Recall that $a_n(t)$'s are computed based on the channel model, feedback, sensing results, and sensing errors. For the next time slot, $a_n(t+1)$ can be estimated as $\hat{a}_n(t+1) = \lambda_n a_n(t) + \mu_n[1 - a_n(t)] = (\lambda_n - \mu_n)a_n(t) + \mu_n$. Recursively, we can derive $\hat{a}_n(t+\tau)$ for the next τ time slots.

$$\hat{a}_n(t+\tau) = (\lambda_n - \mu_n)^\tau a_n(t) + \mu_n \frac{1 - (\lambda_n - \mu_n)^\tau}{1 - (\lambda_n - \mu_n)}. \tag{2.15}$$

At the beginning section of a GoP window, all the base layers will be first transmitted. We start the estimation after all the base layers have been successfully received (possibly with retransmissions). The number of available tiles in the following T_{est} time slots can be estimated as

$$T_e(t) = \sum_{n=1}^{N} \sum_{\tau=0}^{t_{\min}} \hat{a}_n(t+\tau),$$

where

$$\begin{cases} \hat{a}_n(t+0) = a_n(t) \\ t_{\min} = \min\{T_{\text{est}} - 1, T_{\text{GoP}} - (t \bmod T_{\text{GoP}})\}. \end{cases} \tag{2.16}$$

Note that $T_e(t)$ may not be an integer, but it does not affect the outcome of GRD2.

We then adjust the $l_{g,m}$'s based on $T_e(t)$ and $N_{ack}(t)$, the number of ACKs received in time slot t. If $T_e(t) + N_{\text{ack}}(t-1) > T_e(t-1) + N_{\text{ack}}(t-2)$, there are more tiles that can be allocated and we can increase some of the $l_{g,m}$'s. On the other hand, if $T_e(t) + N_{\text{ack}}(t-1) < T_e(t-1) + N_{\text{ack}}(t-2)$, we have to reduce some of the $l_{g,m}$'s. Due to layered videos, when we increase the number of allocated tiles, we only need to consider $l_{g,m}$ for $m = m', m'+1, \ldots, M$, where $\text{MC}_{m'}$ is the highest MC scheme

Algorithm 3: Refined Greedy Algorithm (GRD2) for Each Time Slot ([1], ©2010 IEEE)

Initialize $l_{g,m} = 0$ for all g and m, $A = \{1, 2, \ldots, G\}$, and $N_{ack}(0) = 0$;
Estimate $T_e(1)$ based on the Markov Chain channel model ;
Use GRD1 to find all $l_{g,m}$'s based on $T_e(1)$;
for $(t = 2 : T_{GoP})$ **do**
    Estimate $T_e(t)$;
    if $(T_e(t) + N_{ack}(t-1) < T_e(t-1) + N_{ack}(t-2))$ **then**
        while $\left(\sum_{g=1}^{G}\sum_{m=1}^{M} l_{g,m} > T_e(t) + N_{ack}(t-2)\right)$ **do**
            Find $l_{\hat{g},\hat{m}}$ that can be reduced by 1:
            $\vec{e}_{\hat{g},\hat{m}} = \arg\min_{\forall g, m \in \{m',\ldots,M\}} \left\{\frac{U(\vec{l}) - U(\vec{l} - \vec{e}_{g,m})}{b_{g,m} + R/T_e}\right\}$;
            $\vec{l} = \vec{l} - \vec{e}_{\hat{g},\hat{m}}$;
            if $(\hat{g} \notin A)$ **then** Add $\hat{g}$ to A ;
        end
    end
    if $(T_e(t) + N_{ack}(t-1) > T_e(t-1) + N_{ack}(t-2))$ **then**
        while $\left(\sum_{g=1}^{G}\sum_{m=1}^{M} l_{g,m} \le T_e(t) + N_{ack}(t-1) \textit{and} A \textit{ is not empty}\right)$ **do**
            Find $l_{\hat{g},\hat{m}}$ that can be increased by 1:
            $\vec{e}_{\hat{g},\hat{m}} = \arg\max_{g \in A, m \in \{m',\ldots,M\}} \left\{\frac{U(\vec{l} + \vec{e}_{g,m}) - U(\vec{l})}{b_{g,m} + R/T_e}\right\}$;
            $\vec{l} = \vec{l} + \vec{e}_{\hat{g},\hat{m}}$;
            if $\left(\sum_m b_{\hat{g},m} l_{\hat{g},m} > \bar{R}_{\hat{g}}^e\right)$ **then**
                $\vec{l} = \vec{l} - \vec{e}_{\hat{g},\hat{m}}$;
                Delete $\hat{g}$ from A ;
            end
        end
    end
    Update $N_{ack}(t-1)$;
end

used in the previous time slot. Similarly, when we reduce the number of allocated tiles, we only need to consider $l_{g,m}$ for $m = m', m'+1, \ldots, M$.

The refined greedy algorithm is given in Algorithm 3. For time slot t, the complexity of GRD2 is $O(MGK)$, where $K = |N_{\text{ack}}(t-1) - N_{\text{ack}}(t-2) + T_e(t) - T_e(t-1)|$. Since $K \ll T_e$, the complexity of GRD2 is much lower than GRD1, suitable for execution in each time slot.

2.2.3 *Tile Scheduling in a Time Slot*

In each time slot t, we need to schedule the remaining tiles for transmission on the N channels. Define $\text{Inc}(g, m, i)$ as the increase in the group utility function $U(g)$ after

Algorithm 4: Algorithm for Tile Scheduling in a Time Slot ([1], ©2010 IEEE)

Initialize m_g to the lowest MC that has not been ACKed for all g ;
Initialize i_g to the first packet that has not been ACKed for all g ;
Sort $\{c_n(t)\}$ in decreasing order. Let the sorted channel list be indexed by j ;
for $(j = 1 : N)$ **do**
  Find the group having the maximum increase in $U(g)$: $\hat{g} = \arg\max_{\forall g} \text{Inc}(g, m_g, i_g)$;
  Allocate the tile on channel j to group $\hat{g}$;
  Update $m_{\hat{g}}$ and $i_{\hat{g}}$;
end

the i-th tile in the sub-layer using MC_m is successfully decoded. It can be shown that

$$\text{Inc}(g, m, i) = \sum_{k=m}^{M} (n_{g,k} - n_{g,k+1}) \times \log\left[1 + \frac{\beta_g b_{g,m}}{Q_g^b + \beta_g \sum_{u=1}^{m-1} b_{g,u} l_{g,u} + (i-1)\beta_g b_{g,m}}\right],$$

where $\text{Inc}(g, m, i)$ can be interpreted as the *reward* if the tile is successfully received.

Letting $c_n(t)$ be the probability that the tile is successfully received, then we have $c_n(t) = p_n^{\text{tr}}(t) a_n(t)$. Our objective of tile scheduling is to maximize the expected reward, i.e.,

$$\text{maximize:} \quad \text{E}[\text{Reward}(\vec{\xi})] = \sum_{n=1}^{N} c_n(t) \cdot \text{Inc}(\xi_n), \tag{2.17}$$

where $\vec{\xi} = \{\xi_n\}_{n=1,\ldots,N}$ and ξ_n is the tile allocation for channel n, i.e., representing the three-tuple $\{g, m, i\}$. The TSA algorithm is shown in Algorithm 4, which solves the above optimization problem. The complexity of TSA is $O(N \log N)$. We have the following theorem for TSA.

Theorem 2.2 *E[Reward] is maximized if* $\text{Inc}(\xi_i) > \text{Inc}(\xi_j)$ *when* $c_i(t) > c_j(t)$ *for all* i *and* j.

Proof Suppose there exists a pair of i and j where $\text{Inc}(\xi_i) > \text{Inc}(\xi_j)$ and $c_i(t) < c_j(t)$. We can further increase E[Reward] by switching the tile assignment, i.e., assign channel i to ξ_j and channel j to ξ_i. With this new assignment, the net increase in E[Reward] is

$$\begin{aligned} & c_j(t)\text{Inc}(\xi_i) + c_i(t)\text{Inc}(\xi_j) - c_i(t)\text{Inc}(\xi_i) - c_j(t)\text{Inc}(\xi_j) \\ &= [c_j(t) - c_i(t)][\text{Inc}(\xi_i) - \text{Inc}(\xi_j)] \\ &> 0. \end{aligned}$$

Therefore E[Reward] is maximized when the $\{\text{Inc}(\xi_i)\}$ and $\{c_i(t)\}$ are in the same order. □

2.3 Simulation Results

We evaluate the performance of the proposed CR video multicast framework using a customized simulator implemented with a combination of C and MATLAB. Specifically, the LPs are solved using the MATLAB Optimization Toolbox and the remaining parts are written in C. For the results reported in this section, we have $N = 12$ channels (unless otherwise specified). The channel parameters λ_n and μ_n are set between (0, 1). The maximum allowed collision probability γ_n is set to 0.2 for all the channels unless otherwise specified.

The CR base station multicasts three Common Intermediate Format (CIF, 352 × 288) video sequences to three multicast groups, i.e., *Bus* to group 1, *Foreman* to group 2, and *Mother & Daughter* to group 3. The $n_{1,m}$'s are $\{42, 40, 36, 30, 22, 12\}$ (i.e., 42 users can decode MC_1 signal, 40 users can decode MC_2 signal, and so forth); the $n_{2,m}$'s are $\{51, 46, 40, 32, 23, 12\}$ and the $n_{3,m}$'s are $\{49, 44, 40, 32, 24, 13\}$. The number of bits carried in one tile using the MC schemes are 1, 1.5, 2, 3, 5.3, and 6 kb/s, respectively. We choose T_{GoP}=150 and $T_{\text{est}} = 10$, sensing interval $W = 3$, false alarm probability $\epsilon_n = 0.3$ and miss detection probability $\delta_n = 0.25$ for all n, unless otherwise specified.

In every simulation, we compare three schemes:

- A simple heuristic scheme that equally allocates tiles to each group (Equal Allocation)
- A scheme based on SF (Sequential Fixing)
- A scheme based on the greedy algorithm GRD2 (Greedy Algorithm).

These schemes have increasing complexity in the order of Equal Allocation, Greedy Algorithm,, and Sequential Fixing. They differ on how to solve Problem OPT-Part, while the same tile scheduling algorithm and opportunistic spectrum access scheme are used in all the schemes. Each point in the figures is the average of 10 simulation runs, with 95 % confidence intervals plotted. We observe that the 95 % confidence intervals for Equal Allocation and Greedy Algorithm are negligible, while the 95 % confidence intervals for Sequential Fixing are relatively larger. The C/MATLAB code is executed in a Dell Precision Workstation 390 with an Intel Core 2 Duo E6300 CPU working at 1.86 GHz and a 1,066 MB memory. For a number of channels ranging from N=3 to N=15, the execution times of Equal Allocation and Greedy Algorithm are about a few milliseconds, while Sequential Fixing takes about 2 s.

In Fig. 2.3 we plot the average PSNR among all users in each multicast group. For all the groups, Greedy Algorithm achieves the best performance, with up to 4.2 dB improvements over Equal Allocation and up to 0.6 dB improvements over Sequential Fixing. We find Sequential Fixing achieves a lower PSNR than Equal Allocation for

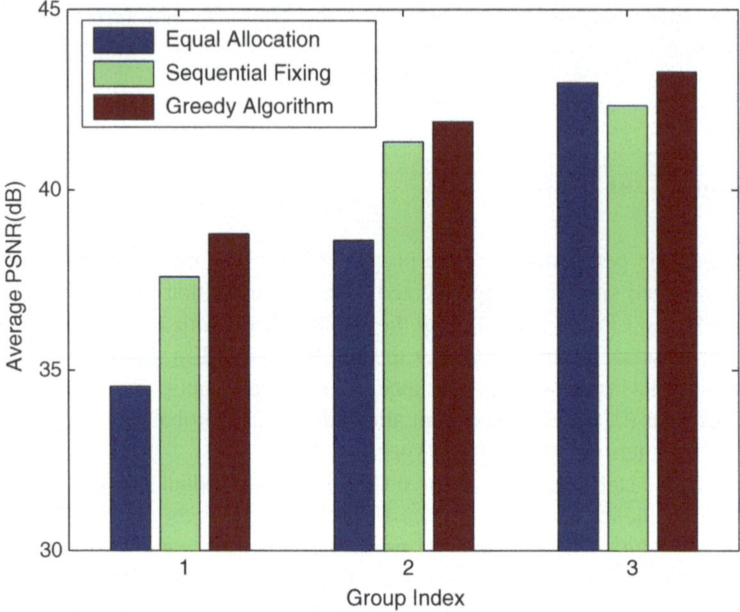

Fig. 2.3 Average PSNR of all multicast users ([1], ©2010 IEEE)

Fig. 2.4 The original (the *right* one) and decoded Frame 53 (the *left* one) at user 1 in group 1 ([1], ©2010 IEEE)

group 3, but higher PSNRs for groups 1 and 2. This is because Equal Allocation does not consider channel conditions and fairness. It achieves better performance for group 3 at the cost of much lower PSNRs for groups 1 and 2. We also plot Frame 53 from the original *Bus* sequence and the decoded video at user 1 of group 1 in Fig 2.4. We choose this user since it is one of the users with lowest PSNR values. The average PSNR of this user is 29.54 dB, while the average PSNR of all group

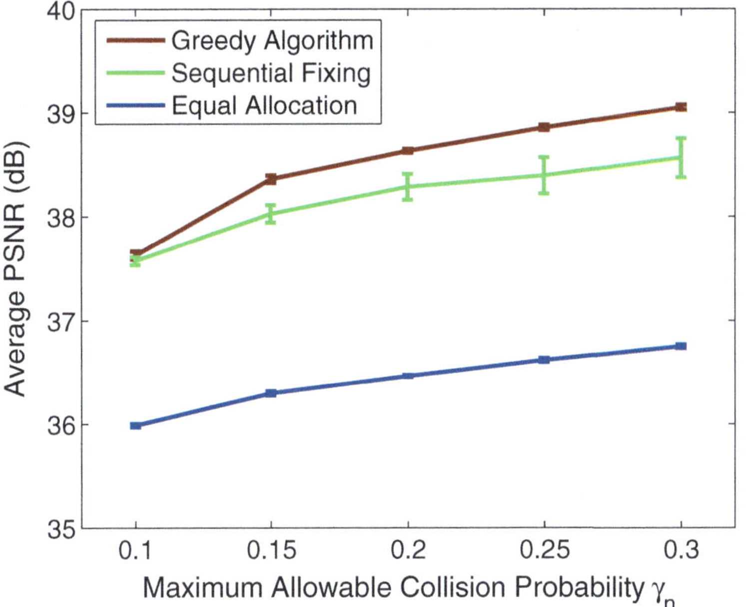

Fig. 2.5 Average PSNR of all users versus γ_n (with 95 % confidence intervals) ([1], ©2010 IEEE)

1 users is 34.6 dB. Compared to the original frame (right), the reconstructed frame (left) looks quite good, although some details are lost.

In Fig. 2.5, we examine the impact of the maximum allowed collision probability γ_n. We increase γ_n from 0.1 to 0.3, and plot the average PSNR values among all the users. When γ_n gets larger, there will be higher chance of collision for the video packets, which hurts the received video quality. However, a higher γ_n also allows a higher transmission probability $p_n^{\text{tr}}(t)$ for the base station (see (2.6)), thus allowing the base station to grab more spectrum opportunities and achieve a higher video rate. The net effect of these two contradicting effects is improved video quality for the range of γ_n values considered in this simulation. This is illustrated in the figure where all the three curves increase as γ_n gets larger. We also observe that the curves for Sequential Fixing and Equal Allocation are roughly parallel to each other, while the Greedy Algorithm curve has a steeper slope. This indicates that Greedy Algorithm is more efficient in exploiting the additional bandwidth allowed by an increased γ_n.

In Fig. 2.6, we examine the impact of number of channels N. We increase N from 3 to 15 in steps of 3, and plot the average PSNR values of all multicast users. As expected, the more the channels, the more the spectrum opportunities for the CR networks, and the better the video quality. Again, we observe that the Greedy Algorithm curve has the steepest slope, implying it is more efficient in exploiting the increased spectrum opportunity for video transmissions.

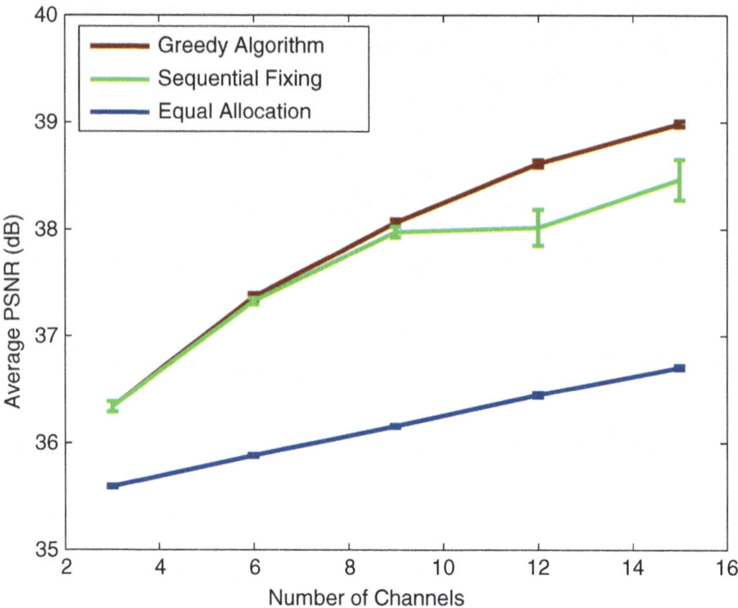

Fig. 2.6 Average PSNR of all users versus N (with 95 % confidence intervals) ([1], ©2010 IEEE)

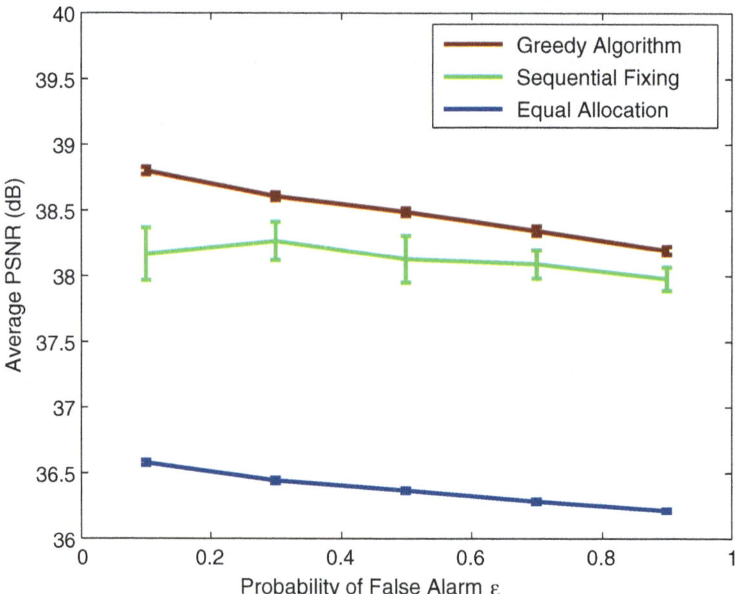

Fig. 2.7 Average PSNR of all users for various $\{\epsilon_n, \delta_n\}$ values (with 95 % confidence intervals) ([1], ©2010 IEEE)

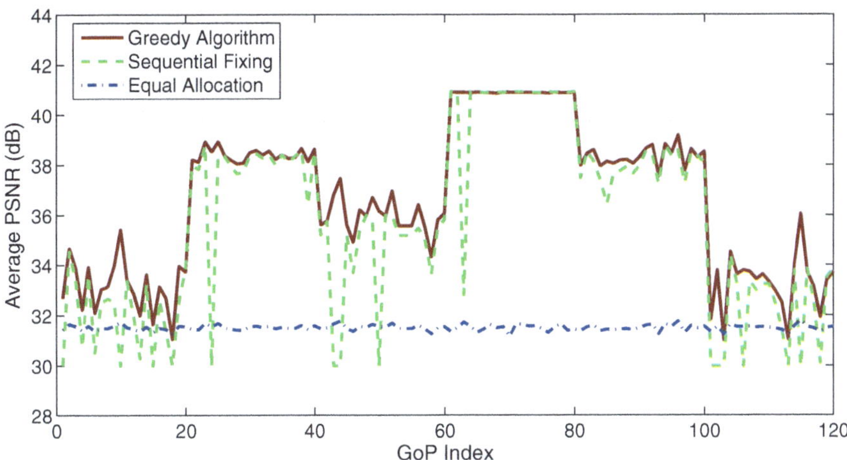

Fig. 2.8 GoP average PSNRs of a tagged user in group 1, when its channel condition varies over time ([1], ©2010 IEEE)

We demonstrate the impact of sensing errors in Fig. 2.7. We test five sets of $\{\epsilon_n, \delta_n\}$ values as follows: {0.10, 0.38}, {0.30, 0.25}, {0.5, 0.17}, {0.70, 0.10} and {0.9, 0.04} [23], and plot the average PSNR values of all users. It is quite interesting to see that the video quality is not very sensitive to sensing errors. Even as ϵ_n is increased nine times from 10 to 90 %, there is only 0.58 dB reduction (or a 1.5 % normalized reduction) in average PSNR when Greedy Algorithm is used. The same can be observed for the other two curves. We conjecture that this is due to the opportunistic spectrum access approach adopted in all the three schemes. A special strength of the proposed approach is that it explicitly considers both types of sensing errors and mitigates the impact of both sensing errors. For example, when the false alarm rate is very high, the base station will not trust the sensing results and will access the channel relatively more aggressively, thus mitigating the negative effect of the high false alarm rate.

Finally, we demonstrate the impact of user channel variations (i.e., due to mobility). We choose a tagged user in group 1 and assume that its channel condition changes every 20 GoPs. The highest MC scheme that the tagged user can decode is changed according to the following sequence: MC3, MC5, MC4, MC6, MC5, and MC3. All other parameters remain the same as in the previous experiments. In Fig. 2.8, we plot the average PSNRs for each GoP at this user that are obtained using the three algorithms. We observe that both Greedy Algorithm and Sequential Fixing can quickly adapt to changing channel conditions. Both algorithms achieve received video qualities commensurate with the channel quality of the tagged user. We also find the video quality achieved by Greedy Algorithm is more stable than that of Sequential Fixing, while the latter curve has some deep fades from time to time. This is due to the fact that Greedy Algorithm has a proven optimality bound, while Sequential Fixing does not provide any guarantee. The Equal Allocation curve is relative constant for the entire period since it does not adapt to channel variations. Although being simple, it does not provide good video quality in this case.

For optimization-driven multimedia systems, there is a trade-off between (i) grabbing all the available resource to maximize media quality and (ii) be less adaptive to network dynamics for a smooth playout. The main objective of this chapter is to demonstrate the feasibility and layout the framework for video streaming over infrastructure-based CR networks, using an objective function of maximizing the overall user utility. We will investigate the interesting problem of trading off resource utilization and smoothness in our future work.

2.4 Related Work

As observed [12, 13], the mainstream CR research has been focused on spectrum sensing and dynamic spectrum access issues. For example, the approach of iteratively sensing a selected subset of available channels has been adopted in the design of CR MAC protocols (e.g., see [18]). The important trade-off between the two types of sensing errors is addressed in-depth in [23].

The equally important QoS issue has been considered only in a few papers [28-34], where the focus is on the so-called network-centric metrics such as maximum throughput and delay [31]. In [34], a game-theoretic framework is described for resource allocation for multimedia transmissions in spectrum agile wireless networks. In this interesting work, each wireless station plays a resource management game, which is coordinated by a network moderator. A mechanism-based resource management scheme determines the amount of transmission time to be allocated to various users on different frequency bands such that certain global system metrics are optimized.

Video multicast, as one of the most important multimedia services, has attracted considerable efforts from the research community. Layered video multicast has been studied in the context of mobile ad hoc networks (e.g., see [35, 36]) and infrastructure-based wireless networks (e.g., see [24, 37]). A greedy algorithm is proposed in [24] for layered video multicast in WiMAX networks with a proven optimality gap $(1-e^{-1/2})$. The proposed GRD1 algorithm extends the work in [24] for FGS video under dynamic channel availability processes. The main difference between this and the prior studies is that, unlike the prior work where the spectrum is exclusively used by video sessions, we need to consider the presence and protection of primary users, which makes the problem more interesting and challenging.

2.5 Conclusions

In this chapter, we investigated the problem of multicasting FGS video in CR networks. The problem formulation took video quality and proportional fairness as objectives, while considering cross-layer design factors such as FGS coding, spectrum sensing, opportunistic spectrum access,primary user protection, scheduling, error control, and modulation. Effective optimization and scheduling algorithms were developed for highly competitive solutions, with proven complexity and optimality bound for the proposed greedy algorithm. Simulation results demonstrated not only the viability of video over CR networks, but also the efficacy of the proposed approach.

Chapter 3
Video over Cooperative CR Networks

As discussed, Cognitive radios (CR) provide an effective solution to meeting the spectrum deficit problem, by exploiting co-deployed networks and sharing underutilized spectrum [12, 38]. On the other hand, *cooperative communications* represent another effective solution to the capacity problem, where wireless nodes help each other in information delivery to achieve the so-called *cooperative diversity* [39, 40].

Wireless relays have been incorporated in the recent IEEE 802.16j (WiMAX) standard and will be included in the next generation IEEE 802.16m standard. Recently, researchers have been exploring the idea of combining these two techniques [2, 41–43], and the potential of cooperative CR networks has been demonstrated with a testbed implementation [41]. Furthermore, Guan et al. in [44] addressed the challenging problem of joint video encoding rate control, power control, relay selection, and channel in cognitive ad hoc networks with cooperative relays. A solution algorithm based on a combination of the branch and bound framework and convex relaxation techniques was proposed, and the performance evaluation results demonstrated the efficacy of cooperation and CR for video streaming.

In this chapter, we investigate relay-assisted multiuser video streaming in a CR cellular networks, where videos are supported to make the best use of the enhanced network capacity. Consider a base station (BS) and multiple relay nodes (RN) that collaboratively stream multiple videos to CR users within the network. It has been shown that the performance of a cooperative relay link is mainly limited by two factors: (i) the *half-duplex operation*, since the BS–RN and the RN–user transmissions cannot be scheduled simultaneously on the same channel [39] and (ii) the *bottleneck channel*, which is usually the BS–user and/or the RN–user channel, which usually has poor quality due to obstacles, attenuation, multipath propagation, and mobility [40]. To support high-quality video service in such a challenging environment, we assume a well-planned relay network where the RNs are connected to the BS with high-speed wireline links [45]. Therefore, the video packets will be available at both the BS and the RNs before their scheduled transmission time, thus allowing advanced cooperative transmission techniques to be adopted for streaming videos.

S. Mao, *Video over Cognitive Radio Networks*, DOI: 10.1007/978-1-4614-4957-7_3,

We focus on the bottleneck channel problem. In particular, we consider zero-forcing, an interference alignment technique where the BS and RNs simultaneously transmit encoded mixed signals to all CR users, such that undesired signals will be canceled and the desired signal can be decoded at each CR user [46–48]. In [49], such cooperative sender-side techniques are termed *interference alignment*, while receiver-side techniques that use overheard (or exchanged via a wireline link) packets to cancel interference is termed *interference cancellation*. Interference alignment is underpinned by recent advances in information theory [47, 48], and the practical implications are addressed and proof-of-concept implementations are reported [47–51].

We first present a stochastic programming formulation of the problem of zero-forcing for video streaming in cooperative CR networks. The cross-layer optimization formulation takes into account important design factors including spectrum sensing, opportunistic spectrum access, cooperative relay, interference alignment, and video QoS requirements. A reformulation of the problem is then presented based on Linear Algebra theory [52], such that the number of variables and computational complexity can be greatly reduced.

We develop effective solution algorithms for the reformulated problem. In particular, three scenarios are considered. In the case of a single licensed channel, we develop a distributed algorithm based on dual decomposition [53], and prove its guaranteed convergence and bounded convergence rate. In the case of multiple licensed channels with channel bonding (where a transmitter can aggregate all the available channels to transmit the encoded signal [54, 55]), we show that the distributed algorithm can still be used to achieve optimal solutions. Finally, in the case of multiple licensed channels without channel bonding, we develop a greedy algorithm that leverages the single channel algorithm for near-optimal solutions, and prove a lower bound for its performance. The proposed algorithms are evaluated with simulations, and are shown to outperform two heuristic schemes that do not incorporate interference alignment with considerable gains.

The remainder of this chapter is organized as follows. The system model and preliminaries are presented in Sect. 3.1. The problem statement is given in Sect. 3.2. The proposed solution algorithms are introduced and analyzed in Sect. 3.3. We present simulation results in Sect. 3.4 and discuss related work in Sect. 3.5. Section 3.6 concludes this chapter.

3.1 Network Model and Assumptions

The cooperative CR network is shown in Fig. 3.1. There is a CR BS (indexed 1) and $(K-1)$ CR RNs (indexed from 2 to K) deployed in the area to serve N active CR users. Let $\mathcal{U} = \{1, 2, \ldots, N\}$ denote the set of active CR users. Assume the BS and all the RNs are equipped with multiple transceivers: one is tuned to the common control channel and the others are used to sense multiple licensed channels at the beginning of each time slot, and to transmit encoded signals to CR users.

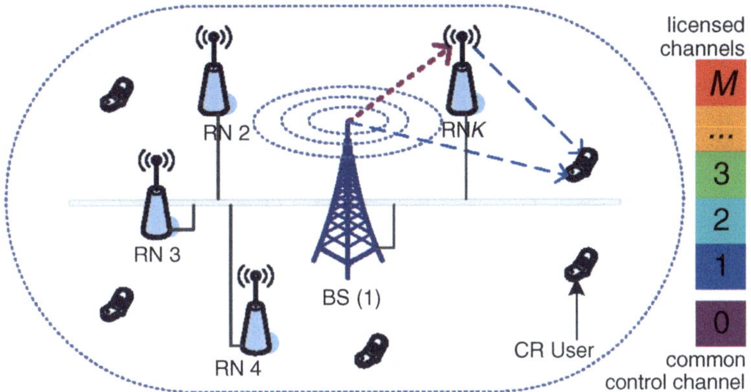

Fig. 3.1 Illustration of the cooperative CR network ([2], © 2012 IEEE)

We consider the case where each CR user has one software defined radio (SDR)-based transceiver, which can be tuned to operate on any of the $(M + 1)$ channels. If the channel bonding/aggregation techniques are used [54, 55], a transmitter can collectively use all the available channels and a CR user can receive from all the available channels, simultaneously. Otherwise, only one licensed channel will be used by a transmitter and a CR user can only receive from a single chosen channel at a time.

Consider the three channels in a traditional cooperative relay link. Usually the BS and RNs are mounted on high towers, and the BS–RN channel has good quality due to line-of-sight (LOS) communications and absence of mobility. On the other hand, a CR user is typically on the ground level. The BS–user and RN–user channels usually have much poorer quality due to obstacles, attenuation, multipath propagation, and mobility. To support high-quality video service, we assume a well-planned relay network, where the RNs are connected to the BS via broadband wireline connections (e.g., as in femtocell networks [3]). Alternatively, free space optical links can also be used to provide multigigabit rates between the BS and the RNs [56]. As a result, the video packets will always be available for transmission (with suitable channel coding and retransmission) at the RNs at their scheduled transmission time. To cope with the much poorer BS–user and RN–user channels, the BS and RNs adopt interference alignment to cooperatively transmit video packets to CR users, while exploiting the spectrum opportunities in the licensed channels.

3.1.1 Spectrum Access

The BS and the RNs sense the licensed channels and exchange their sensing results over the common control channel during the sensing phase. Given L sensing results obtained for channel m, the corresponding sensing result vector is

$\vec{\Theta}_L^m = [\Theta_1^m, \Theta_2^m, \ldots, \Theta_L^m]$. Let $P_m^A(\vec{\Theta}_L^m) := P_m^A(\Theta_1^m, \Theta_2^m, \ldots, \Theta_L^m)$ be the conditional probability that channel m is available, which can be computed iteratively as shown in our prior work [3]:

$$\begin{aligned}
P_m^A(\Theta_1^m) &= \left[1 + \frac{\eta_m}{1-\eta_m} \times \frac{(\delta_1^m)^{1-\Theta_1^m}(1-\delta_1^m)^{\Theta_1^m}}{(\epsilon_1^m)^{\Theta_1^m}(1-\epsilon_1^m)^{1-\Theta_1^m}}\right]^{-1} \\
P_m^A(\vec{\Theta}_l^m) &:= P_m^A(\Theta_1^m, \Theta_2^m, \ldots, \Theta_l^m) \\
&= \left\{1 + \left[\frac{1}{P_m^A(\Theta_1^m, \Theta_2^m, \ldots, \Theta_{l-1}^m)} - 1\right] \times \right. \\
&\qquad \left. \frac{(\delta_l^m)^{1-\Theta_l^m}(1-\delta_l^m)^{\Theta_l^m}}{(\epsilon_l^m)^{\Theta_l^m}(1-\epsilon_l^m)^{1-\Theta_l^m}}\right\}^{-1}, l \geq 2.
\end{aligned}$$

For each channel m, define an index variable $D_m(t)$ for the BS or RNs to access the channel in time slot t. That is,

$$D_m(t) = \begin{cases} 0, & \text{access channel } m \text{ in time slot } t \\ 1, & \text{otherwise,} \end{cases} \quad m = 1, 2, \ldots, M. \tag{3.1}$$

With sensing result $P_m^A(\vec{\Theta}_L^m)$, each channel m will be opportunistically accessed. Let the probability be $P_m^D(\vec{\Theta}_L^m)$ that channel m will be accessed in time slot t (i.e., when $D_m(t) = 0$). The optimal channel access probability can be computed as

$$P_m^D(\vec{\Theta}_L^m) = \min\left\{\gamma_m / \left[1 - P_m^A(\vec{\Theta}_L^m)\right], 1\right\}. \tag{3.2}$$

Let $\mathcal{A}(t)$ be the set of available channels in time slot t. It follows that $\mathcal{A}(t) := \{m \mid D_m(t) = 0\}$.

3.1.2 Zero-Forcing Precoding

We next briefly describe the main idea of zero-forcing precoding considered in this chapter. Interested readers are referred to [49, 51] for insightful examples, a classification of various interference alignment scenarios, and practical considerations.

Consider two transmitters (denoted as s_1 and s_2) and two receivers (denoted as d_1 and d_2). Let X_1 and X_2 be the signals corresponding to the packets to be sent to d_1 and d_2, respectively. With interference alignment, the transmitters s_1 and s_2 send compound signals $a_{1,1}X_1 + a_{1,2}X_2$ and $a_{2,1}X_1 + a_{2,2}X_2$, respectively, to the two receivers d_1 and d_2, simultaneously. If channel noise is ignored, the received signals Y_1 and Y_2 can be written as

$$\begin{bmatrix} Y_1 \\ Y_2 \end{bmatrix} = \begin{bmatrix} G_{1,1} & G_{1,2} \\ G_{2,1} & G_{2,2} \end{bmatrix}^{\mathsf{T}} \begin{bmatrix} a_{1,1} & a_{1,2} \\ a_{2,1} & a_{2,2} \end{bmatrix} \begin{bmatrix} X_1 \\ X_2 \end{bmatrix}$$
$$:= \mathbf{G}^{\mathsf{T}} \times \mathbf{A} \times \vec{X}, \tag{3.3}$$

where $G_{i,j}$ is the channel gain from transmitter s_i to receiver d_j.

From (3.3), it can be seen that both receivers can perfectly decode their signals if the transformation matrix $\mathbf{A}$ is chosen to be $\{\mathbf{G}^{\mathsf{T}}\}^{-1}$, i.e., the inverse of the channel gain matrix. With this technique, the transmitters are able to send packets simultaneously and the interference between the two concurrent transmissions can be effectively canceled at both receivers [49].

3.2 Problem Formulation

We formulate the problem of interference alignment for scalable video streaming over cooperative CR networks in this section. As discussed in Sect. 3.1, the video packets are available at both the BS and all the RNs before their scheduled transmission time; the BS and RNs adopt interference alignment to overcome the poor BS–CR user and RN–CR user channels.

Let X_j denote the signal to be transmitted to user j, which has unit power. As illustrated in Sect. 3.1.2, transmitter k sends a compound signal $\sum_{j \in \mathcal{U}} a_{k,j} X_j$ to all active CR users, where $a_{k,j}$'s are the weights to be determined. Ignoring channel noise, we can compute the received signal Y_n at a user n as

$$\begin{aligned} Y_n &= \sum_{k=1}^{K} G_{k,n} \sum_{j=1}^{N} a_{k,j} X_j \\ &= \sum_{k=1}^{K} \sum_{j=1}^{N} a_{k,j} G_{k,n} X_j \\ &= \sum_{j=1}^{N} X_j \sum_{k=1}^{K} a_{k,j} G_{k,n}, \quad n = 1, 2, \ldots, N, \end{aligned} \tag{3.4}$$

where $G_{k,n}$ is the channel gain from the BS (i.e., $k = 1$) or an RN k to user n. For user n, only signal X_n should be decoded and the coefficients of all other signals should be forced to zero. The *zero-forcing constraints* can be written as

$$\sum_{k=1}^{K} a_{k,j} G_{k,n} = 0, \quad \text{for all } j \neq n. \tag{3.5}$$

Usually the total transmit power of the BS and every RN is limited by a peak power $P_{\max}$. Since X_j has unit power, for all j, the power of each transmitted signal is the square sum of all the coefficients $a_{k,j}^2$. The *peak power constraint* can be written as

$$\sum_{j=1}^{N} |a_{k,j}|^2 \leq P_{\max}, \quad k = 1, \ldots, K. \tag{3.6}$$

Recall that each CR user has one SDR transceiver that can be tuned to receive from any of the $(M+1)$ channels, when channel bonding is not adopted. Let b_j^m be a binary variable indicating that user j selects licensed channel m. It is defined as

$$b_j^m = \begin{cases} 1, & \text{if user } n \text{ receives from channel } m \\ 0, & \text{otherwise,} \end{cases} \\ j = 1, \ldots, N, \; m = 1, \ldots, M. \tag{3.7}$$

Then, we have the following *transceiver constraint*:

$$\sum_{m \in \mathcal{A}(t)} b_j^m \leq 1, \quad j = 1, \ldots, N. \tag{3.8}$$

After introducing the channel selection variables b_j^m's, the overall channel gain becomes

$$G_{k,j} = \sum_{m \in \mathcal{A}(t)} b_j^m H_{k,j}^m, \tag{3.9}$$

where $H_{k,j}^m$ is the channel gain from the BS (i.e., $k = 1$) or an RN k to user j on channel m.

Let w_j^t be the PSNR of user j's reconstructed video at the beginning of time slot t and W_j^t the PSNR of user j's reconstructed video at the end of time slot t. In time slot t, w_j^t is already known, while W_j^t is a random variable depending on the resource allocation and primary user activity during the time slot. That is, w_j^{t+1} is a realization of W_j^t.

The quality of reconstructed MGS video can be modeled with a linear equation [57]:

$$W(R) = \alpha + \beta \times R, \tag{3.10}$$

where $W(R)$ is the average peak signal-to-noise ratio (PSNR) of the reconstructed MGS video, R is the average data rate, and α and β are constants depending on the specific video sequence and codec.

We formulate a multistage stochastic programming problem to maximize the sum of expected logarithm of the PSNR's at the end of the GOP, i.e.,

$$\sum_{j=1}^{N} \mathbb{E}\left[\log(W_j^T)\right],$$

for proportional fairness among the video sessions [25]. It can be shown that the multistage stochastic programming problem can be decomposed into T serial sub-problems, one for each time slot t, as [4]:

$$\text{maximize: } \sum_{j=1}^{N} \mathbb{E}\left[\log(W_j^t)|w_j^t\right] \tag{3.11}$$

subject to:

$$W_j^t = w_j^t + \Psi_j^t \tag{3.12}$$

$$b_j^m \in \{0, 1\},\ a_{k,j} \geq 0,\ \text{for all } m, j, k \tag{3.13}$$

Constraints (3.5), (3.6) and (3.8),

where Ψ_j^t is a random variable that depends on spectrum sensing, power allocation, and channel selection in time slot t. This is a mixed integer nonlinear programming problem (MINLP), with binary variables b_j^m's and continuous real variables $a_{k,j}$'s.

In particular, Ψ_j^t can have two possible values: (i) zero, if the packet is not successfully received due to collision with primary users; (ii) the PSNR increase achieved in time slot t if the packet is successfully received, denoted as λ_j^t. The PSNR increase can be computed as

$$\lambda_j^t = \frac{\beta_j B}{T} \log_2\left(1 + \frac{1}{N_0}\left(\sum_{k=1}^{K} a_{k,j} G_{k,j}\right)^2\right), \tag{3.14}$$

where N_0 is the noise power and B is the channel bandwidth.

User j can successfully receive a video packet from channel m if it tunes to channel m (i.e., $b_j^m = 1$) and the BS and RNs transmit on channel m (i.e., with probability $P_m^D(\vec{\Theta}_L^m)$). The probability that user j successfully receives a video packet, denoted as P_j^t, is

$$P_j^t = \sum_{m \in \mathcal{A}(t)} b_j^m P_m^D(\vec{\Theta}_L^m). \tag{3.15}$$

Therefore, we can expand the expectation in (3.11) to obtain a reformulated problem:

$$\text{maximize: } \sum_{j=1}^{N} \mathbb{E}\left[P_j^t \log(w_j^t + \lambda_j^t) + (1 - P_j^t)\log(w_j^t)\right] \tag{3.16}$$

subject to:

Constraints (3.5), (3.6), (3.8), and (3.13).

3.3 Solution Algorithms

In this section, we develop effective solution algorithms to the stochastic programming problem (3.11). In Sect. 3.3.1, we first consider the case of a single licensed channel, and derive a distributed, optimal algorithm with guaranteed convergence and bounded convergence speed. We then address the case of multiple licensed channels. If channel bonding/aggregation techniques are used [54, 55], the distributed algorithm in Sect. 3.3.1 can still be applied to achieve optimal solutions. We finally consider the case of multiple licensed channels without channel bonding, and develop a greedy algorithm with a performance lower bound in Sect. 3.3.3.

3.3.1 Case of a Single Channel

Property

Consider the case when there is only one licensed channel, i.e., when $M = 1$. The K transmitters, including the BS and $(K - 1)$ RNs, send video packets to active users using the licensed channel when it is sensed idle.

Definition 3.1. *A set of vectors is linearly independent if none of them can be written as a linear combination of the other vectors in the set* [52].

For user j, the weight and channel gain vectors are:

$$\vec{a}_j = [a_{1,j}, a_{2,j}, \ldots, a_{K,j}]^{\mathsf{T}}$$

and

$$\vec{G}_j = [G_{1,j}, G_{2,j}, \ldots, G_{K,j}]^{\mathsf{T}},$$

where T denotes *matrix transpose*. Due to spatial diversity, we assume that the $\vec{G}_j$ vectors are linearly independent [47].

Lemma 3.1. *To successfully decode each signal X_j, $j = 1, 2, \ldots, N$, the number of active users N should be smaller than or equal to the number of transmitters K.*

Proof. From (3.5), it can be seen that $\vec{a}_j$ is orthogonal to the $(N - 1)$ vectors $\vec{G}_n$'s, for $n \neq j$. Since $\vec{a}_j$ is a K by 1 vector, there are at most $(K - 1)$ vectors that are orthogonal to $\vec{a}_j$. Since the $\vec{G}_j$ vectors are linearly independent, it follows that $(N - 1) \leq (K - 1)$ and therefore $N \leq K$. □

According to Lemma 3.1, the following additional constraints should be enforced for the channel selection variables.

$$\sum_{j=1}^{N} b_j^m \leq K, \quad \text{for all } m \in \mathcal{A}(t). \tag{3.17}$$

That is, the number of active users receiving from any channel m cannot be more than the number of transmitters on that channel, which is K in the single channel case and less than or equal to K in the multiple channels case. We first assume that N is not greater than K, and will remove this assumption in the following section.

Reformulation and Complexity Reduction

With a single channel, all active users receive from channel 1. Therefore $b_j^1 = 1$, and $b_j^m = 0$, for $m > 1$, $j = 1, 2, \ldots, N$. The formulated problem is now reduced to a nonlinear programming problem with constraints (3.5), (3.6), and (3.13). If the number of active users is $N = 1$, the solution is straightforward: all the transmitters send the same signal X_1 to the single user using their maximum transmit power $P_{\max}$.

In general, the reduced problem can be solved with the dual decomposition technique [53] (i.e., a primal dual algorithm). This problem has $K \times N$ primal variables (i.e., the $a_{k,j}$'s), and we need to define $N(N-1)$ dual variables (or, Lagrangian Multipliers) for constraints (3.5) and K dual variables for constraints (3.6). These numbers could be large for even moderate-sized systems. Before presenting the solution algorithm, we first derive a reformulation of the original problem (3.16) that can greatly reduce the number of primal and dual variables, such that the computational complexity can be reduced.

Lemma 3.2. *Each vector $\vec{a}_j = [a_{1,j}, a_{2,j}, \ldots, a_{K,j}]^{\mathsf{T}}$ can be represented by the linear combination of r nonzero, linearly independent vectors, where $r = K - N + 1$.*

Proof. From (3.5), each vector $\vec{a}_j$ is orthogonal to $\vec{G}_i$ where $j \neq i$. Define a reduced matrix $\mathbf{G}_{-j}$ obtained by deleting $\vec{G}_j$ from $\mathbf{G}$, i.e.,

$$\mathbf{G}_{-j} = [\vec{G}_1, \ldots, \vec{G}_{j-1}, \vec{G}_{j+1}, \ldots, \vec{G}_N].$$

Then $\vec{a}_j$ is a solution to the homogeneous linear system $\mathbf{G}_{-j}^{\mathsf{T}} \vec{x} = 0$. Since, we assume that the $\vec{G}_i$'s are all linearly independent, the columns of $\mathbf{G}_{-j}$ are also linearly independent [52]. Thus, the rank of $\mathbf{G}_{-j}$ is $(N-1)$. The solution belongs to the null space of $\mathbf{G}_{-j}$. The dimension of the null space is $r = K - (N-1)$ according to the Rank-nullity Theorem [52]. Therefore, each $\vec{a}_j$ can be presented by the linear combination of r linearly independent vectors. □

Let $\mathbf{e}_j = \{\vec{e}_{j,1}, \vec{e}_{j,2}, \ldots, \vec{e}_{j,r}\}$ be a *basis* for the null space of $\mathbf{G}_{-j}$. There are many methods to obtain the basis, such as Gaussian Elimination. However, we show that it is not necessary to solve the homogeneous linear system of equations $\mathbf{G}_{-j}^{\mathsf{T}} \vec{x} = 0$ to get the basis for every different j value. Therefore, the computational complexity can be further reduced.

Our algorithm for computing a basis is shown in Algorithm 1. In Steps 1–6, we first solve the homogeneous linear system of equations $\mathbf{G}^{\mathsf{T}} \vec{x} = 0$ to get a basis $[\vec{v}_1, \vec{v}_2, \ldots, \vec{v}_{K-N}]$. Note that if K is equal to N, the basis is the empty set $\emptyset$. We

Algorithm 1: Basis Computation Algorithm ([2], © 2012 IEEE)

1 **if** $(K > N)$ **then**
2   Solve homogeneous linear system $\mathbf{G}^{\mathsf{T}}\vec{x} = 0$ and get basis $[\vec{v}_1, \ldots, \vec{v}_{K-N}]$;
3   **for** $(i = 1 : K - N)$ **do**
4         $\vec{e}_{j,i} = \vec{v}_i$, for all j ;
5   **end**
6 **end**
7 **for** $(j = 1 : N)$ **do**
8   Orthogonalize $\mathbf{G}_{-j}$ and get $(N - 1)$ orthogonal vectors $\vec{w}_{j,i}$'s ;
9   Calculate $\vec{e}_{j,r}$ as in (3.18) ;
10 **end**

then set the $K - N$ basis vectors to be the first $K - N$ vectors in all the basses $\mathbf{e}_j$, $j = 1, 2, \ldots, N$. In Step 8, we orthogonalize each $\mathbf{G}_{-j}$ and obtain $(N - 1)$ orthogonal vectors $\vec{\omega}_{j,i}$, $i = 1, 2, \ldots, N - 1$. Finally in Step 9, we let the rth vector $\vec{e}_{j,r}$ be orthogonal to all the $\vec{\omega}_{j,i}$'s by subtracting all the projections on each $\vec{\omega}_{j,i}$ from $\vec{G}_j$ (recall that $r = K - N + 1$). The operation is:

$$\begin{aligned} \vec{e}_{j,r} &= \vec{e}_{j,N-K+1} \\ &= \vec{G}_j - \sum_{i=1}^{N-1} \frac{\vec{G}_j^{\mathsf{T}}\vec{\omega}_{j,i}}{\vec{\omega}_{j,i}^{\mathsf{T}}\vec{\omega}_{j,i}}\vec{\omega}_{j,i}. \end{aligned} \tag{3.18}$$

Lemma 3.3. *The solution space constructed by the basis* $[\vec{v}_1, \vec{v}_2, \ldots, \vec{v}_{K-N}]$ *is a sub-space of the solution space of* $\mathbf{G}_{-j}^{\mathsf{T}}\vec{x} = 0$ *for all* j.

Proof. It is easy to see that each vector $\vec{v}_i$ is a solution of $\mathbf{G}_{-j}^{\mathsf{T}}\vec{x} = 0$ by substituting $\vec{x}$ with $\vec{v}_i$, for $i = 1, 2, \ldots, K - N$. □

Lemma 3.4. *The vectors* $[\vec{v}_1, \vec{v}_2, \ldots, \vec{v}_{K-N}, \vec{e}_{j,r}]$ *computed in Algorithm 1 is a basis of the null space of* $\mathbf{G}_{-j}$.

Proof. Obviously, the $\vec{v}_i$'s are linearly independent. From (3.18), it is easy to verify that $\vec{e}_{j,r}$ is orthogonal to all the $\vec{\omega}_{j,i}$'s. Therefore, $\vec{e}_{j,r}$ is also a solution to system $\mathbf{G}_{-j}^{\mathsf{T}}\vec{x} = 0$. Since $\vec{G}_j$ and $\vec{\omega}_{j,i}$ are orthogonal to all the $\vec{v}_i$'s, and $\vec{e}_{j,r}$ is a linear combination of $\vec{G}_j$ and $\vec{\omega}_{j,i}$, $\vec{e}_{j,r}$ is also orthogonal and linearly independent to all the $\vec{v}_i$'s. The conclusion follows. □

Define coefficients $\vec{c}_j = [c_{j,1}, c_{j,2}, \ldots, c_{j,r}]^{\mathsf{T}}$. Then we can represent $\vec{a}_j$ as a linear combination of the basis vectors, i.e., $\vec{a}_j = \sum_{l=1}^{r} c_{j,l}\vec{e}_{j,l} = \mathbf{e}_j\vec{c}_j$. Equation (3.14) can be rewritten as

Table 3.1 Comparison of Computational Complexity ([2], © 2012 IEEE)

	Original problem	Reformulated problem
Primal variables	KN	$(K-N+1)N$
Dual variables	$N(N-1)+K$	K

$$\begin{aligned}\lambda_j^t &= \frac{\beta_j B}{T}\log_2\left(1+\frac{1}{N_0}\left(\vec{c}_j^{\mathsf{T}}\mathbf{e}_j^{\mathsf{T}}\vec{G}_j\right)^2\right)\\ &= \frac{\beta_j B}{T}\log_2\left(1+\frac{1}{N_0}\left(c_{j,r}\vec{e}_{j,r}^{\mathsf{T}}\vec{G}_j\right)^2\right). \qquad (3.19)\end{aligned}$$

The second equality is because the first $K-N$ column vectors in $\mathbf{e}_j$ are orthogonal to G_j. The random variable W_j^t in the objective function now only depends on $c_{j,r}$. The peak power constraint can be revised as

$$\sum_{j=1}^{N}[\mathbf{e}_j(k)\vec{c}_j]^2 \le P_{\max}, \quad k=1,\ldots,K, \qquad (3.20)$$

where $\mathbf{e}_j(k)$ is the kth row of matrix $\mathbf{e}_j$.

With such a reformulation, the number of primal and dual variables can be greatly reduced. In Table 3.1, we show the numbers of variables in the original problem and in the reformulated problem. The number of primary variables is reduced from KN to $(K-N+1)N$, and the number of dual variables is reduced from $N(N-1)+K$ to K. Such reductions result in greatly reduced computational complexity.

Distributed Algorithm

To solve the reformulated problem, we define non-negative dual variables $\vec{\mu} = [\mu_1,\ldots,\mu_K]^{\mathsf{T}}$ for the inequality constraints. The Lagrangian function is

$$\begin{aligned}\mathcal{L}(\mathbf{c},\vec{\mu}) &= \sum_{j=1}^{N}\mathbb{E}\left[\log(W_j^t(c_{j,r}))|w_j^t\right] + \sum_{k=1}^{K}\mu_k(P_{\max} - \sum_{j=1}^{N}[\mathbf{e}_j(k)\vec{c}_j]^2)\\ &= \sum_{j=1}^{N}\mathcal{L}_j(\vec{c}_j,\vec{\mu}) + P_{\max}\sum_{k=1}^{K}\mu_k, \qquad (3.21)\end{aligned}$$

where $\mathbf{c}$ is a matrix consisting of all column vector $\vec{c}_j$'s and

$$\mathcal{L}_j(\vec{c}_j,\vec{\mu}) = \mathbb{E}\left[\log(W_j^t(c_{j,r}))|w_j^t\right] - \sum_{k=1}^{K}\mu_k[\mathbf{e}_j(k)\vec{c}_j]^2.$$

The above problem can be decomposed into N subproblems and solved iteratively [53]. In every Step $\tau \geq 1$, for given vector $\vec{\mu}(\tau)$, each CR user solves the following subproblem using local information

$$\vec{c}_j(\tau) = \arg\max \mathcal{L}_j(\vec{c}_j, \vec{\mu}(\tau)). \tag{3.22}$$

Obviously, the objective function in (3.22) is concave. Therefore, there is a unique optimal solution. The CR users then exchange their solutions over the common control channel. To solve the primal problem, we adopt the gradient method [53].

$$\vec{c}_j(\tau + 1) = \vec{c}_j(\tau) + \phi \nabla \mathcal{L}_j(\vec{c}_j(\tau), \vec{\mu}(\tau)), \tag{3.23}$$

where $\nabla \mathcal{L}_j(\vec{c}_j(\tau), \vec{\mu}(\tau))$ is the gradient of the primal problem and ϕ is a small positive step size.

The master dual problem for a given $\mathbf{c}(\tau)$ is:

$$\min_{\mu_i \geq 0, i=1,\ldots,K} q(\vec{\mu}) = \sum_{j=1}^{N} \mathcal{L}_j(\vec{c}_j(\tau), \vec{\mu}) + P_{\max} \sum_{k=1}^{K} \mu_k. \tag{3.24}$$

Since the Lagrangian function is differentiable, the subgradient iteration method can be adopted.

$$\vec{\mu}(\tau + 1) = [\vec{\mu}(\tau) - \rho(\tau)\vec{g}(\tau)]^+, \tag{3.25}$$

where $\rho(\tau) = \frac{q(\vec{\mu}(\tau)) - q(\vec{\mu}^*)}{||\vec{g}(\tau)||^2}$ is a positive step size, $\vec{\mu}^*$ is the optimal solution, $\vec{g}(\tau) = \nabla q(\vec{\mu}(\tau))$ is the gradient of the dual problem, and $[\cdot]^+$ denotes the projection onto the nonnegative axis. Since the optimal solution $\vec{\mu}^*$ is unknown a priori, we choose the mean of the objective values of the primal and dual problems as an estimate for $\vec{\mu}^*$ in the algorithm. The updated $\mu_k(\tau+1)$ will again be used to solve the subproblems (3.22). Since the problem is convex, we have strong duality; the duality gap between the primal and dual problems will be zero. The distributed algorithm is shown in Algorithm 2, where $0 \leq \kappa \ll 1$ is a threshold for convergence.

Performance Analysis

We analyze the performance of the distributed algorithm in this section. In particular, we prove that it converges to the optimal solution at a speed faster than $\sqrt{1/\tau}$ as τ goes to infinity.

Theorem 3.1. *The series $q(\vec{\mu}(\tau))$ converges to $q(\vec{\mu}^*)$ as τ goes to infinity and the square error sum $\sum_{\tau=1}^{\infty}(q(\vec{\mu}(\tau)) - q(\vec{\mu}^*))^2$ is bounded.*

Proof. For the optimality gap, we have:

Algorithm 2: Algorithm for the Case of a Single Channel ([2], © 2012 IEEE)

if $(N = 1)$ **then**
  Set $a_{k,j}$ to $P_{\max}$ for all k ;
else
  Set $\tau = 0$; $\vec{\mu}(0)$ to positive values ;
  Set $\mathbf{c}(0)$ to random values ;
  Compute bases $\mathbf{e}_j$'s as in Algorithm 1 ;
  repeat
    $\tau = \tau + 1$;
    Compute $\vec{c}_j(\tau)$ as in (3.23) ;
    Broadcast $\vec{c}_j(\tau)$ on the common control channel ;
    Update $\vec{\mu}(\tau)$ as in (3.25) ;
  until $(||\vec{\mu}(\tau) - \vec{\mu}(\tau-1)|| > \kappa)$;
  Compute $a_{k,j}$'s ;
end

$$
\begin{aligned}
&||\vec{\mu}(\tau+1) - \vec{\mu}^*||^2 = ||[\vec{\mu}(\tau) - \rho(\tau)\vec{g}(\tau)]^+ - \vec{\mu}^*||^2 \\
&\le ||\vec{\mu}(\tau) - \rho(\tau)\vec{g}(\tau) - \vec{\mu}^*||^2 \\
&= ||\vec{\mu}(\tau) - \vec{\mu}^*||^2 - 2\rho(\tau)(\vec{\mu}(\tau) - \vec{\mu}^*)^{\mathsf{T}}\vec{g}(\tau) + (\rho(\tau))^2||\vec{g}(\tau)||^2 \\
&= ||\vec{\mu}(\tau) - \vec{\mu}^*||^2 - 2\rho(\tau)(q(\vec{\mu}(\tau)) - q(\vec{\mu}^*)) + (\rho(\tau))^2||\vec{g}(\tau)||^2.
\end{aligned}
$$

Since the step size is $\rho(\tau) = \frac{q(\vec{\mu}(\tau)) - q(\vec{\mu}^*)}{||\vec{g}(\tau)||^2}$, it follows that

$$
\begin{aligned}
||\vec{\mu}(\tau+1) - \vec{\mu}^*||^2 &\le ||\vec{\mu}(\tau) - \vec{\mu}^*||^2 - \frac{(q(\vec{\mu}(\tau)) - q(\vec{\mu}^*))^2}{||\vec{g}(\tau)||^2} \\
&\le ||\vec{\mu}(\tau) - \vec{\mu}^*||^2 - \frac{(q(\vec{\mu}(\tau)) - q(\vec{\mu}^*))^2}{\hat{g}^2}, \qquad (3.26)
\end{aligned}
$$

where $\hat{g}^2$ is an upper bound of $||\vec{g}(\tau)||^2$. Since the second term on the right-hand side of (3.26) is nonnegative, it follows that $\lim_{\tau\to\infty} q(\vec{\mu}(\tau)) = q(\vec{\mu}^*)$.

Summing inequality (3.26) over τ, we have

$$
\sum_{\tau=1}^{\infty}(q(\vec{\mu}(\tau)) - q(\vec{\mu}^*))^2 \le \hat{g}^2||\vec{\mu}(1) - \vec{\mu}^*||^2.
$$

That is, the square error sum is upper bounded. □

Theorem 3.2. *The sequence $\{q(\vec{\mu}(\tau))\}$ converges faster than $\{1/\sqrt{\tau}\}$ as τ goes to infinity.*

Proof. Assume $\lim_{\tau\to\infty}\sqrt{\tau}(q(\vec{\mu}(\tau)) - q(\vec{\mu}^*)) > 0$. Then there is a sufficiently large τ' and a positive number ξ such that $\sqrt{\tau}(q(\vec{\mu}(\tau)) - q(\vec{\mu}^*)) \ge \xi$, for all $\tau \ge \tau'$. Taking the square sum from τ' to ∞, we have:

$$\sum_{\tau=\tau'}^{\infty}(q(\vec{\mu}(\tau)) - q(\vec{\mu}^*))^2 \geq \xi^2 \sum_{\tau=\tau'}^{\infty} \frac{1}{\tau} = \infty. \tag{3.27}$$

Equation (3.27) contradicts with Theorem 3.1, which states that the square error sum $\sum_{\tau=1}^{\infty}(q(\vec{\mu}(\tau)) - q(\vec{\mu}^*))^2$ is bounded. Therefore, we have

$$\lim_{\tau\to\infty} \frac{q(\vec{\mu}(\tau)) - q(\vec{\mu}^*)}{1/\sqrt{\tau}} = 0, \tag{3.28}$$

indicating that the convergence speed of $q(\vec{\mu}(\tau))$ is faster than that of $1/\sqrt{\tau}$. □

3.3.2 Case of Multiple Channels with Channel Bonding

When there are multiple licensed channels, we first consider the case when the channel bonding and aggregation techniques are used by the transmitters and CR users [54, 55]. With channel bonding, a transmitter can utilize all the available channels in $\mathcal{A}(t)$ collectively to transmit the mixed signal. We assume that at the end of the sensing phase in each time slot, CR users tune their SDR transceiver to the common control channel to receive the set of available channels $\mathcal{A}(t)$ from the BS. Then each CR user can receive from all the channels in $\mathcal{A}(t)$ and decode its desired signal from the compound signal it receives.

This case is similar to the case of a single licensed channel. Now all the active CR users receive from the set of available channels $\mathcal{A}(t)$. We thus have $b_j^m = 1$, for $m \in \mathcal{A}(t)$, and $b_j^m = 0$, for $m \notin \mathcal{A}(t)$, $j = 1, 2, \ldots, N$. When all the b_j^m's are determined this way, problem (3.11) is reduced to a nonlinear programming problem with constraints (3.5) and (3.6). The distributed algorithm described in Sect. 3.3.1 can be applied to solve this reduced problem to get optimal solutions.

3.3.3 Case of Multiple Channels Without Channel Bonding

We finally consider the case of multiple channels without channel bonding, where each CR user has a narrow band SDR transceiver and can only receive from one of the channels. We first present a greedy algorithm that leverages the optimal algorithm in Algorithm 2 for near-optimal solutions, and then derive a lower bound for its performance.

Greedy Algorithm

When $M > 1$, the optimal solution to problem (3.11) depends also on the binary variables b_j^m's, which determines whether user j receives from channel m. Recall

Algorithm 3: Channel Selection Algorithm for the Case of Multiple Channels without Channel Bonding ([2], © 2012 IEEE)

1 Initialize $\vec{b}$ to a zero vector, user set $\mathcal{U} = \{1, \ldots, N\}$ and user-channel set $\mathcal{C} = \mathcal{U} \times \mathcal{A}(t)$;
2 **while** ($\mathcal{C} \neq \emptyset$) **do**
3 Find the user-channel pair $\{j', m'\}$, such that $\{j', m'\} = \arg\max_{\{(j,m) \in \mathcal{C}\}} \{\Phi(\vec{b} + \vec{v}_j^m) - \Phi(\vec{b})\}$;
4 Set $\vec{b} = \vec{b} + \vec{v}_{j'}^{m'}$ and remove j' from $\mathcal{U}$;
5 **if** $\left(\sum_{j=1}^{N} b_j^{m'} = K\right)$ **then**
6 Remove m' from $\mathcal{A}(t)$;
7 **end**
8 Update user-channel set $\mathcal{C} = \mathcal{U} \times \mathcal{A}(t)$;
9 **end**

that there are two constraints for the b_j^m's: (i) each user can use at most one channel (see (3.8)); (ii) the number of users on the same channel cannot exceed the number of transmitters K (see (3.17)). Let $\vec{b}$ be the channel allocation vector with elements b_j^m's, and $\Phi(\vec{b})$ the corresponding objective value for a given user-channel allocation $\vec{b}$.

We take a two-step approach to solve problem (3.11). First, we apply the greedy algorithm in Algorithm 3 to choose one available channel in $\mathcal{A}(t)$ for each CR user (i.e., to determine $\vec{b}$). Second, we apply the algorithm in Algorithm 2 to obtain a near-optimal solution for the given channel allocation $\vec{b}$.

In Algorithm 3, $\vec{v}_j^m$ is a unit vector with 1 for the $[(j-1) \times M + m]$th element and 0 for all other elements, and $\vec{b} = \vec{b} + \vec{v}_{j'}^{m'}$ indicates choosing channel m' for user j'. In each iteration, the user-channel pair (j', m') that can achieve the largest increase in the objective value is chosen, as in Step 3. The complexity of the greedy algorithm in the worst case is $O(K^2 M^2)$.

Performance Bound

We next analyze the greedy algorithm and derive a lower bound for its performance. Let ν_l be the sequence from the first to the lth user-channel pair selected by the greedy algorithm. The increase in objective value is denoted as

$$F_l := F(\nu_l, \nu_{l-1}) = \Phi(\nu_l) - \Phi(\nu_{l-1}). \tag{3.29}$$

Sum up (3.29) from 1 to L. We have $\sum_{l=1}^{L} F_l = \Phi(\nu_L)$ since $\Phi(\nu_0) = 0$. Let Ω be the global optimal solution for user-channel allocation. Define π_l as a subset of Ω. For given ν_l, π_l is the subset of user-channel pairs that cannot be allocated due to the conflict with the lth user-channel allocation ν_l (but not conflict with the user-channel allocations in ν_{l-1}).

Lemma 3.5. *Assume the greedy algorithm stops in L steps, we have*

$$\Phi(\Omega) \leq \Phi(\nu_L) + \sum_{l=1}^{L} \sum_{\sigma \in \pi_l} F(\sigma \cup \nu_{l-1}, \nu_{l-1}).$$

Proof. The proof is similar to the proof of Lemma 7 in [3] and is omitted for brevity. □

Theorem 3.3. *The greedy algorithm for channel selection in Algorithm 3 can achieve an objective value that is at least* $\frac{1}{|\mathcal{A}(t)|}$ *of the global optimum in each time slot.*

Proof. According to Lemma 3.5, it follows that:

$$\begin{aligned} \Phi(\Omega) &\leq \Phi(\nu_L) + \sum_{l=1}^{L} |\pi_l| F_l \\ &\leq \Phi(\nu_L) + (|\mathcal{A}(t)| - 1) \sum_{l=1}^{L} F_l \\ &= |\mathcal{A}(t)| \Phi(\nu_L). \end{aligned} \tag{3.30}$$

The second inequality is due to the fact that each user can choose at most one channel and there are at most $(|\mathcal{A}(t)| - 1)$ pairs in π_l according to the definition. The equality in (3.30) is because $\sum_{l=1}^{L} F_l = \Phi(\nu_L)$. Then we have:

$$\frac{1}{|\mathcal{A}(t)|} \Phi(\Omega) \leq \Phi(\nu_L) \leq \Phi(\Omega). \tag{3.31}$$

The greedy heuristic solution is lower bounded by $1/|\mathcal{A}(t)|$ of the global optimum. □

Define competitive ratio $\chi = \Phi(\nu_L)/\Phi(\Omega) = 1/|\mathcal{A}(t)|$. Assume all the licensed channels have identical utilization η. Since $|\mathcal{A}(t)|$ is a random variable, we take the expectation of χ and obtain:

$$\mathbb{E}[\chi] = \eta^M + \sum_{n=1}^{M} \left(\frac{1}{n}\right) \eta^{M-n} (1 - \eta)^n. \tag{3.32}$$

In Fig. 3.2, we evaluate the impact of channel utilization η and the number of licensed channels M on the competitive ratio. We increase η from 0.05 to 0.95 in steps of 0.05 and increase M from 6 to 12 in steps of 2. The lower bound (3.31) becomes tighter when η is larger or when M is smaller. For example, when $\eta = 0.6$ and $M = 6$, the greedy algorithm solution is guaranteed to be no less than 52.7%

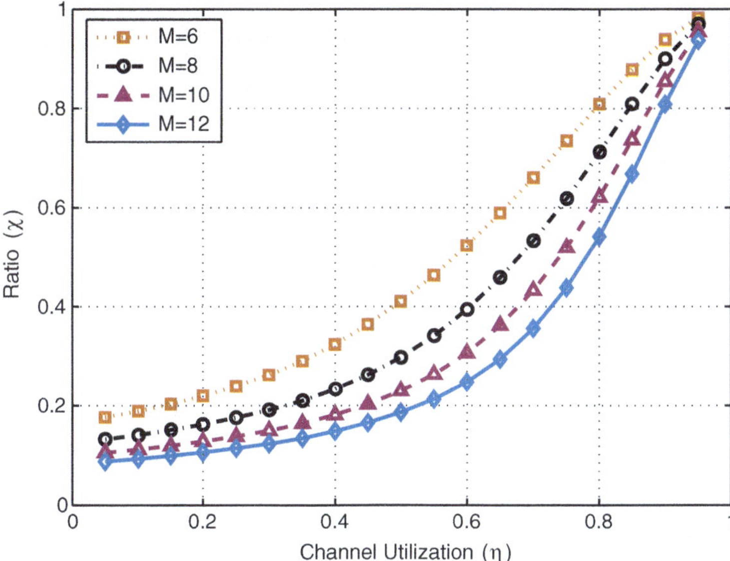

Fig. 3.2 Competitive ratio $\mathbb{E}[\chi]$ defined in (3.32) versus channel utilization η ([2], © 2012 IEEE)

of the global optimal. when η is increased to 0.95, the greedy algorithm solution is guaranteed to be no less than 98.3% of the global optimal.

3.4 Performance Evaluation

We evaluate the performance of the proposed algorithms with a MATLAB implementation and the JVSM 9.13 Video Codec. We present simulation results for the following two scenarios: (i) a single licensed channel and (ii) multiple licensed channels without channel bonding, since we observe similar performance for the case of multiple licensed channels with channel bonding. For comparison purpose, we also developed two simpler heuristic schemes that do not incorporate interference alignment.

- *Heuristic 1*: each CR user selects the best channel in $\mathcal{A}(t)$ based on channel condition. The time slot is equally divided among the active users receiving from the same channel, to send their signals separately in each time slice.
- *Heuristic 2*: in each time slot, the active user with the best channel is selected for each available channel. The entire time slot is used to transmit this user's signal.

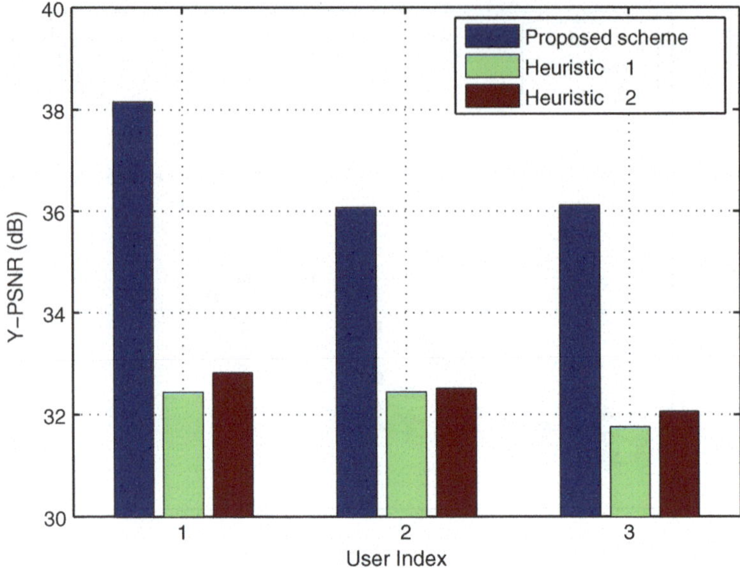

Fig. 3.3 Received video quality for each CR user with a single channel ([2], © 2012 IEEE)

3.4.1 Case of a Single Licensed Channel

In the first scenario, there are $K = 4$ transmitters, i.e., one BS and three RNs. The channel utilization η is set to 0.6 and the maximum allowable collision probability γ is set to 0.2. There are three active CR users, each receives an MGS video stream from the BS: *Bus* to CR user 1, *Mobile* to CR user 2, and *Harbor* to CR user 3. The video sequences are in the Common Intermediate Format (CIF, 252×288). The GOP size of the videos is 16 and the delivery deadline T is 10. The false alarm probability is $\epsilon_l^m = 0.3$ and the miss detection probability is $\delta_l^m = 0.3$ for all spectrum sensors. The channel bandwidth B is 1 MHz. The peak power limit is 10 W for all the transmitters, unless otherwise specified.

We first plot the average Y-PSNRs of the three reconstructed MGS videos in Fig. 3.3, i.e., only the Y (Luminance) component of the original and reconstructed videos are used. Among three schemes, the proposed algorithm achieves the highest PSNR value, while the two heuristic algorithms have similar performance. Note that the proposed algorithm is optimal in the single channel case. It achieves significant improvements ranging from 3.1 to 5.25 dB over the two heuristic algorithms. Such PSNR gains are considerable, since in video coding and communications, a half dB gain is distinguishable and worth pursing.

We next examine the convergence rate of the distributed algorithm. According to Theorem 3.2, the distributed algorithm converges at a speed faster than $1/\sqrt{\tau}$ asymptotically. We compare the optimality gap of the proposed algorithm, i.e., $|q(\tau) - q^*|$, with series $10/\sqrt{\tau}$ in Fig. 3.4. Both curves converge to 0 as τ goes to infinity.

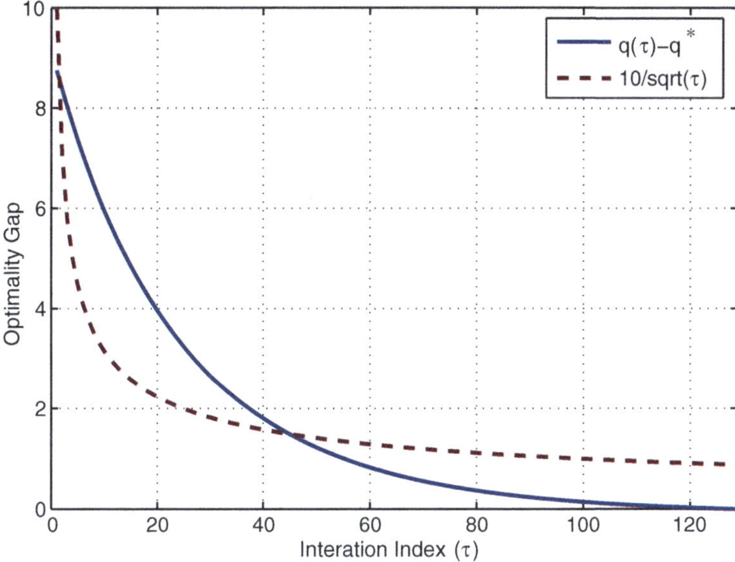

Fig. 3.4 Convergence rate of the distributed algorithm with a single channel ([2], © 2012 IEEE)

It can be seen that the convergence speed, i.e., the slope of the curve of the proposed scheme is larger than that of $10/\sqrt{\tau}$ after about 10 iterations. The convergence of the optimality gap is much faster than $10/\sqrt{\tau}$, which exhibits a heavy tail.

In the case of multiple channels with channel bonding, the performance of the proposed algorithm is similar to that in the single channel case. We omit the results for lack of space.

3.4.2 Case of Multiple Channels Without Channel Bonding

We next investigate the second scenario with six licensed channels and four transmitters. There are 12 CR users, each streaming one of the three different videos *Bus*, *Mobile*, and *Harbor*. The rest of the parameters are the same as those in the single channel case, unless otherwise specified. Equation (3.30) can also be interpreted as an upper bound on the global optimal, i.e., $\Phi(\Omega) \leq |\mathcal{A}(t)|\Phi(\nu_L)$, which is also plotted in the figures. Each point in the following figures is the average of 10 simulation runs with different random seeds. The 95 % confidence intervals are plotted as error bars, which are generally negligible.

The impact of channel utilization η on received video quality is presented in Fig. 3.5. We increase η from 0.3 to 0.9 in steps of 0.15, and plot the Y-PSNRs of reconstructed videos averaged over all the 12 CR users. Intuitively, a smaller η allows more transmission opportunities for CR users, thus allowing the CR users to achieve

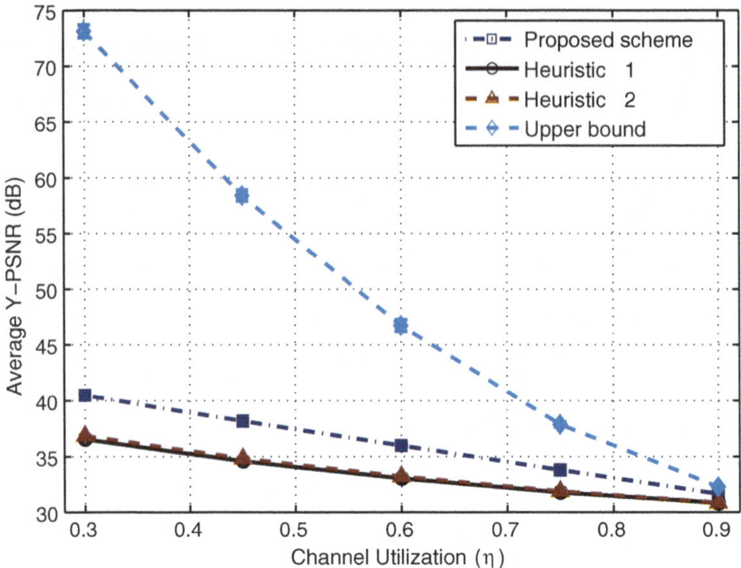

Fig. 3.5 Reconstructed video quality versus channel utilization η in the multichannel without channel bonding case ([2], © 2012 IEEE)

higher video rates and better video quality. This is shown in the figure, in which all four curves decrease as η is increased. We also observe that the gap between the upper bound and proposed schemes becomes smaller as η gets larger, from 32.65 dB when $\eta = 0.3$ to 0.63 dB when $\eta = 0.9$. This trend is also shown in Fig. 3.2. The proposed scheme outperforms the two heuristic schemes with considerable gains, ranging from 0.8 to 3.65 dB.

Finally, we investigate the impact of the number of transmitters K on the video quality. In this simulation we increase K from 2 to 6 with step size 1. The average Y-PSNRs of all the 12 CR users are plotted in Fig. 3.6. As expected, the more transmitters, the more effective the interference alignment technique, and thus the better the video quality. The proposed algorithm achieves gains ranging from 1.78 dB (when $K = 2$) to 4.55 dB (when $K = 6$) over the two heuristic schemes.

3.5 Related Work

This work is closely related to the prior work on cooperative communications [39, 40], where relays are used to achieve cooperative diversity, and that on CR networking [12], where spectrum opportunities in licensed channels are exploited for the benefit of unlicensed users. There have been significant advances in these areas, which laid out the foundation for this work. In particular, researchers have been exploring

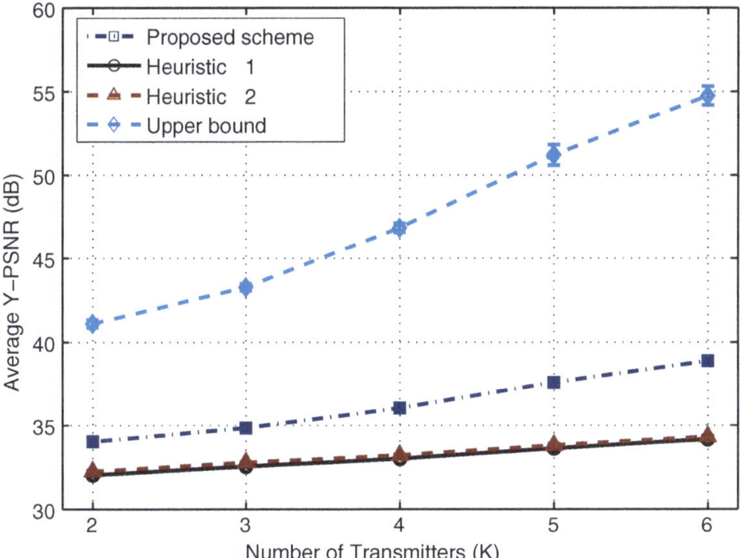

Fig. 3.6 Reconstructed video quality versus number of transmitters K in the multichannel without channel bonding case ([2], © 2012 IEEE)

the idea of combining these two techniques [41, 42]. In [41], an overview of cooperative relay scenarios and related issues was presented, along with a GNU Radio implementation of a MAC protocol. In [42], a centralized heuristic was presented to address the relay selection and spectrum allocation problem in CR networks.

The problem of video over CR networks has only been studied in a few recent papers [1, 3, 4, 58–60]. In [58], a dynamic channel selection scheme was proposed for CR users to transmit videos over multiple channels. In [59], a distributed joint routing and spectrum sharing algorithm for video streaming over CR ad hoc networks was described and evaluated with simulations. In our prior work, we considered video multicast in an infrastructure-based CR network [1], unicast video streaming over multihop CR networks [4] and CR femtocell networks [3]. In [60], the impact of system parameters residing in different network layers are jointly considered to achieve the best possible video quality for CR users. Unlike the heuristic approaches in [58, 59], the analytical and optimization approach taken in this chapter yields algorithms with optimal or bounded performance. The cooperative relay and interference alignment techniques also distinguish this chapter from prior work on this topic.

As point-to-point link capacity approaches the Shannon limit, there has been considerable interest on exploiting interference to improve wireless network capacity [47–51]. In addition to information theoretic work on asymptotic capacity [47, 48], practical issues have been addressed in [49–51]. In [50], the authors presented a practical design of analog network coding to exploit interference and allow concurrent transmissions, which does not make any synchronization assumptions. In [51], interference alignment and cancelation is incorporated in MIMO LANs, and the network capacity is shown, analytically and experimentally, to be almost doubled. In [49], the authors presented a general algorithm for identifying interference alignment and cancelation opportunities in practical multihop mesh networks. The impact of synchronization and channel estimation was evaluated through a GNU Radio implementation. Our work was motivated by these interesting papers, and we incorporate interference alignment in cooperative CR networks and exploit the enhanced capacity for wireless video streaming.

3.6 Conclusions

In this chapter, we investigated the problem of interference alignment for MGS video streaming in a cooperative relay enhanced CR network. We presented a stochastic programming formation, and derived a reformulation that leads to considerable reduction in computational complexity. A distributed optimal algorithm was developed for the case of a single channel and the case of multichannel with channel bonding, with proven convergence and convergence speed. We also presented a greedy algorithm for the multichannel without channel bonding case, with a proven performance bound. The proposed algorithms are evaluated with simulations and are shown to outperform two heuristic schemes without interference alignment with considerable gains.

Chapter 4
Video over Femto CR Networks

Due to the use of open space as transmission medium, capacity of wireless networks is usually limited by interference. When a mobile user moves away from the base station, a considerably larger transmit power is needed to overcome attenuation, while causing interference to other users and deteriorating network capacity. To this end, femtocells provide an effective solution that brings network infrastructure closer to mobile users.

A femtocell is a small (e.g., residential) cellular network, with a *femto base station* (FBS) connected to the owner's broadband wireline network [61, 62]. The FBS serves approved users when they are within the coverage. Among the many benefits, femtocells are shown effective on improving network coverage and capacity [61]. Due to reduced distance, transmit power can be greatly reduced, leading to prolonged battery life, improved signal-to-interference-plus-noise ratio (SINR), and better spatial reuse of spectrum.

Femtocells have received significant interest from the wireless industry. Although highly promising, many important problems should be addressed to fully harvest their potential, such as interference mitigation, resource allocation, synchronization, and QoS provisioning [61, 62]. It is also critical for the success of this technology to support important applications such as real-time video streaming in femtocell networks.

In this chapter, we investigate the problem of video streaming in femtocell cognitive radio (CR) networks. We consider a femtocell network consisting of a *macro base station* (MBS) and multiple FBS's. The femtocell network is co-located with a primary network with multiple licensed channels. The idea is to exploit CR and dynamic spectrum access to utilize spectrum opportunities in the licensed channels for streaming videos [12]. Femtocell subscribers (or, CR users) and FBS's sense licensed channels, and determine which channel(s) and which base station (i.e., the MBS or an FBS) to use for delivering video packets based on sensing results. The objective is to maximize the capacity of the femtocell CR network on carrying real-time video data, while bounding the interference to primary users.

S. Mao, *Video over Cognitive Radio Networks*, DOI: 10.1007/978-1-4614-4957-7_4,

This is a challenging problem due to the stringent QoS requirements of real-time videos and, on the other hand, the new dimensions of network dynamics (i.e., channel availability) and uncertainties (i.e., spectrum sensing and errors) found in CR networks. It also involves a long list of design factors that necessitates cross-layer optimization, such as spectrum sensing and errors, dynamic spectrum access, interference modeling and primary user protection, channel allocation, and video performance, among others.

We adopt Scalable video coding (SVC) in our system. SVC encodes a video into multiple substreams, subsets of which can be decoded to provide different quality levels for the reconstructed video [57]. Such scalability is very useful for video streaming systems, especially in CR networks, to accommodate heterogeneous channel availabilities and dynamic network conditions. We consider H.264/SVC medium grain scalable (MGS) videos, since MGS can achieve better rate-distortion performance over Fine-granularity-scalability (FGS), although it only has Network Abstraction Layer (NAL) unit-based granularity [57].

The unique femtocell network architecture and the scalable video allow us to develop a framework that captures the key design issues and trade-offs, and to formulate a *stochastic programming* problem. It has been shown that the deployment of femtocells has a significant impact on the network performance [61]. In this chapter, we examine three deployment scenarios. In the case of a single FBS, we apply *dual decomposition* to develop a distributed algorithm that can compute the optimal solution. In the case of multiple noninterfering FBS's, we show that the same distributed algorithm can be used to compute optimal solutions. In the case of multiple interfering FBS's, we develop a greedy algorithm that can compute near-optimal solutions, and prove a closed-form lower bound for its performance based on an *interference graph* model. The proposed algorithms are evaluated with simulations, and are shown to outperform three alternative schemes with considerable gains.

The remainder of this chapter is organized as follows. The system model and preliminaries are given in Sect. 4.1. We present the problem formulation and develop solution algorithms in Sect. 4.2. Simulation results are shown in Sect. 4.3. The related work is discussed in Sect. 4.4. Section 4.5 concludes this chapter.

4.1 System Model and Preliminaries

4.1.1 Spectrum and Network Model

Consider a spectrum consisting of $(M+1)$ channels, including one common, unlicensed channel (indexed as channel 0) and M licensed channels (indexed as channels 1 to M). The M licensed channels are allocated to a primary network, and the common channel is exclusively used by all CR users. We assume all the channels follow a synchronized time slot structure [12]. The capacity of each licensed channel is B_1 Mbps, while the capacity of the common channel is B_0 Mbps. The channel states evolve independently, while the occupancy of each licensed channel follows a two-state discrete-time Markov process.

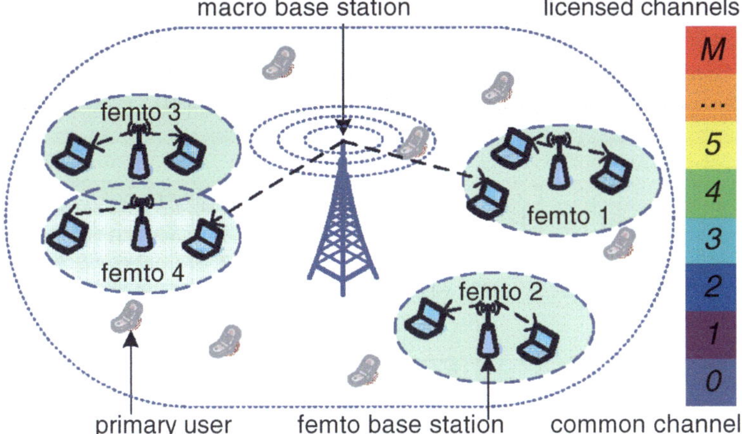

Fig. 4.1 A femtocell CR network with one MBS and four FBS's ([3], © 2012 IEEE)

The femtocell CR network is shown in Fig. 4.1. There is an MBS and N FBS's deployed in the area to serve CR users. The N FBS's are connected to the MBS (and the Internet) via broadband wireline connections. Due to advances in antenna technology, it is possible to equip multiple antennas at the base stations. The MBS has one antenna that is always tuned to the common channel. Each FBS is equipped with multiple antennas (e.g., M) and is able to sense multiple licensed channels at the beginning of each time slot. There are K_i CR users in femtocell i, $i = 1, 2, \ldots, N$, and $\sum_{i=1}^{N} K_i = K$. Each CR user has a software radio transceiver, which can be tuned to any of the $M + 1$ channels. A CR user will either connect to a nearby FBS using one or more of the licensed channels or to the MBS via the common channel.

Although the CR users are mobile, we assume constant topology during a time slot. If the topology is changed during a time slot, the video transmission will only be interrupted for the time slot, since the proposed algorithms are executed in every time slot for new channel assignment and schedule.

4.1.2 Spectrum Sensing and Access

The femtocell CR network is within the coverage of the infrastructure-based primary network. Both FBS's and CR users sense the channels to identify spectrum opportunities in each time slot. Each time slot consists of (i) a *sensing phase*, when CR users and FBS's sense licensed channels, (ii) a *transmission phase*, when CR users and FBS's attempt to access licensed channels, and (iii) an *acknowledgment phase*, when acknowledgments (ACK) are returned to the source.

Cooperative sensing policy is also adopted here. We also adopt a *hypothesis test* to detect channel availability. We assume that each CR user chooses one channel to sense in each time slot, since it only has one transceiver. The sensing results will be shared among CR users and FBS's via the common channel in the sensing phase. Given L sensing results on channel m, the availability of channel m, i.e., $P_m^A(\Theta_1^m, \ldots, \Theta_L^m)$, can be computed iteratively as follows [3].

$$P_m^A(\Theta_1^m) = \left[1 + \frac{\eta_m}{1-\eta_m} \times \frac{(\delta_1^m)^{1-\Theta_1^m}(1-\delta_1^m)^{\Theta_1^m}}{(\epsilon_1^m)^{\Theta_1^m}(1-\epsilon_1^m)^{1-\Theta_1^m}}\right]^{-1} \tag{4.1}$$

$$\begin{aligned} P_m^A(\vec{\Theta}_l^m) &= P_m^A(\Theta_1^m, \Theta_2^m, \ldots, \Theta_l^m) \\ &= \left\{1 + \left[\frac{1}{P_m^A(\Theta_1^m, \Theta_2^m, \ldots, \Theta_{l-1}^m)} - 1\right] \right. \\ &\quad \left. \times \frac{(\delta_l^m)^{1-\Theta_l^m}(1-\delta_l^m)^{\Theta_l^m}}{(\epsilon_l^m)^{\Theta_l^m}(1-\epsilon_l^m)^{1-\Theta_l^m}}\right\}^{-1}, l = 2, \ldots, L. \end{aligned} \tag{4.2}$$

We adopt a probabilistic approach: based on sensing results $\vec{\Theta}_m$, we have $D_m(t) = 0$ with probability $P_m^D(\vec{\Theta}_m)$ and $D_m(t) = 1$ with probability $1 - P_m^D(\vec{\Theta}_m)$. For primary user protection, the collision probability with primary users caused by CR users should be bounded. The probability $P_m^D(\vec{\Theta}_m)$ is determined as follows:

$$P_m^D(\vec{\Theta}_m) = \min\left\{\gamma_m/[1 - P_m^A(\vec{\Theta}_m)], 1\right\}. \tag{4.3}$$

Let $\mathcal{A}(t) := \{m | D_m(t) = 0\}$ be the set of available channels in time slot t. Then $G^t = \sum_{m \in \mathcal{A}(t)} P_m^A(\Theta_1^m)$ is the expected number of available channels. These channels will be accessed in the transmission phase of time slot t.

4.1.3 Channel Model

Without loss of generality, we consider independent block fading channels that is widely used in prior work [63]. The channel fading-gain process is piecewise constant on blocks of one time slot, and fading in different time slots are independent. Let $f_X^{i,j}(x)$ denote the *probability density function* of the received SINR X from a base station i at CR user j. We assume the packet can be successfully decoded if the received SINR exceeds a threshold H. The packet loss probability from base station i to CR user j is:

$$P_{i,j} = \Pr\{X \le H\} = \int_0^H f_X^{i,j}(x)dx = F_X^{i,j}(H), \tag{4.4}$$

where $F_X^{i,j}(H)$ is the cumulative density function of X.

In the case of correlated fading channels, which can be modeled as finite state Markov Process [64], the packet loss probability in the next time slot can be estimated from the known state of the previous time slot and the transition probabilities. If the packet is successfully decoded, the CR user returns an ACK to the base station in the ACK phase. We assume ACKs are always successfully delivered.

4.1.4 Video Performance Measure

We assume each active CR user receives a real-time video stream from either the MSB or an FSB. Without loss of generality, we adopt the MGS option of H.264/SVC, for scalability to accommodate the high variability of network bandwidth in CR networks.

Due to real-time constraint, each Group of Pictures (GOP) of a video stream must be delivered in the next T time slots. With MGS, enhancement layer NAL units can be discarded from a quality scalable bit stream, and thus packet-based quality scalable coding is provided. Our approach is to encode the video according to the maximum rate the channels can support. During transmission, only part of the MGS video gets transmitted as allowed by the current available channel bandwidth. The video packets are transmitted in decreasing order of their significance in decoding. When a truncated MGS video is received and decoded, the PSNR is computed by substituting the effective rate of the received MGS video into (4.5) given below, thus the original video is not required.

Without loss of generality, we assume that the last wireless hop is the bottleneck; video data is available at the MBS and FBS's when they are scheduled to be transmitted. The quality of reconstructed MGS video can be modeled as [57]:

$$W(R) = \alpha + \beta \times R, \tag{4.5}$$

where $W(R)$ is the average peak signal-to-noise ratio (PSNR) of the reconstructed video, R is the received data rate, α and β are constants depending on the video sequence and codec.

We verified (4.5) using an H.264/SVC codec and the *Bus*, *Mobile*, and *Harbor* test sequences. In Fig. 4.2, the markers are obtained by truncating the encoded video's enhancement layer at different positions to obtain different effective rates, while the curves are computed using (4.5). The curves fit well with measurements for the three sequences. It is worth noting that PSNR may not be a good measure of video quality as compared with alternative metrics such as MS-SSIM [65]. The main reason for choosing PSNR is that there is a closed-form model relating it to network level metrics-video rate. With the closed-form model, we can have a mathematical formulation of the scheduling/resource allocation problem, and derive effective algorithms. Should such closed-form models be available for MS-SSIM, it is possible to incorporate it into the optimization framework as well.

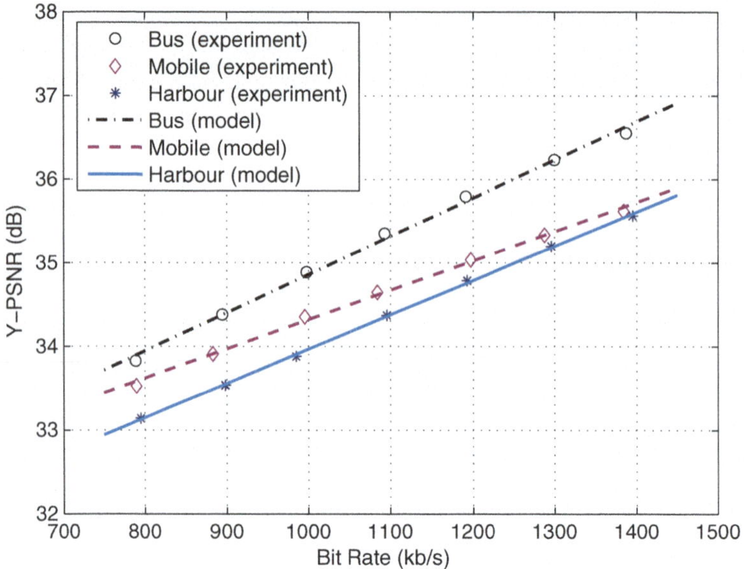

Fig. 4.2 Rate-distortion curves of three H.264/SVC MGS videos ([3], © 2012 IEEE)

4.2 MGS Video over Femtocell CR Networks

In this section, we address the problem of resource allocation for MGS videos over femtocell CR networks. We first examine the case of a single FBS, and then the more general case of multiple noninterfering or interfering FBS's. The algorithms for the single and no-interfering FBS cases are distributed ones and optimal. The algorithm for the interfering FBS case is a centralized one that can be executed at the MBS. To simplify notation, we omit the time slot index t for most of the variables in this Section. For example, x represents a variable for time slot t, x^- represents the variable in time slot $(t-1)$, and x^+ represents the variable in time slot $(t+1)$.

4.2.1 *Case of Single FBS*

Formulation

We first consider the case of a single FBS in the CR network, where the FBS can use all the G available channels to stream videos to K active CR users. Let w_j be the PSNR of CR user j at the beginning of time slot t and W_j the PSNR of CR user j at the end of time slot t. In time slot t, w_j is already known; W_j is a random variable that depends on channel condition and primary user activity; and w_j^+ is a *realization* of W_j. Let $\xi_{0,j}$ and $\xi_{1,j}$ indicate the random packet losses from the MBS and FBS, respectively, to CR user j in time slot t. That is, $\xi_{i,j}$ is 1 with probability $\bar{P}_{i,j} = 1 - P_{i,j}$ and 0 with probability $P_{i,j}$. Due to block fading channels, $P_{i,j}$'s do not change within the time slot.

Let $\rho_{0,j}$ and $\rho_{1,j}$ be the portions of time slot t when CR user j receives video data from the MBS and FBS, respectively. The average PSNR is computed every T time slots. We first have $W_j(0) = \alpha_j$, when $t = 0$. In each time slot t, the CR user receives $\xi_{0,j}\rho_{0,j}B_0$ bits through the MBS, and $\xi_{1,j}\rho_{1,j}GB_1$ bits through the FBS (assuming that OFDM is used), which contribute an increase of $\beta(\xi_{0,j}\rho_{0,j}B_0 + \xi_{1,j}\rho_{1,j}GB_1)/T$ to the total PSNR in this T time slot interval, according to (4.5). Therefore we have the following recursive relationship:

$$\begin{aligned} W_j &= W_j^- + \beta(\xi_{0,j}\rho_{0,j}B_0 + \xi_{1,j}\rho_{1,j}GB_1)/T \\ &= W_j^- + \xi_{0,j}\rho_{0,j}R_{0,j} + \xi_{1,j}\rho_{1,j}GR_{1,j}, \end{aligned} \tag{4.6}$$

where $R_{0,j} = \beta B_0/T$ and $R_{1,j} = \beta B_1/T$.

For proportional fairness, we aim to maximize the sum of the logarithms of the PSNRs of all CR users [25]. We formulate a *multistage stochastic programming problem* by maximizing the *expectation* of the logarithm-sum at time T.

$$\text{maximize:} \quad \sum_{j=1}^{K} \mathbb{E}[\log(W_j(T))] \tag{4.7}$$

subject to:

$$\begin{aligned} W_j &= W_j^- + \xi_{0,j}\rho_{0,j}R_{0,j} + \xi_{1,j}\rho_{1,j}GR_{1,j}, \\ &\qquad\qquad j = 1, \ldots, K, \; t = 1, \ldots, T \\ &\sum_{j=1}^{K} \rho_{i,j} \le 1, \quad i = 0, 1, \; t = 1, \ldots, T \\ &\rho_{i,j} \ge 0, \; i = 0, 1, \; j = 1, \ldots, K, \; t = 1, \ldots, T. \end{aligned}$$

$R_{0,j} = \beta_j B_0/T$ and $R_{1,j} = \beta_j B_1/T$ are constants for the jth MGS video.

At the beginning of the last time slot T, a realization

$$\boldsymbol{\xi}_{[T-1]} = [\vec{\xi}_1, \vec{\xi}_2, \ldots, \vec{\xi}_{T-1}]$$

is known, where

$$\vec{\xi}_t = [\xi_{0,1}^t, \xi_{0,2}^t, \ldots, \xi_{0,K}^t, \xi_{1,1}^t, \ldots, \xi_{1,K}^t], t = 1, 2, \ldots, T - 1.$$

It can be shown that the multistage stochastic programming problem (4.7) can be decomposed into T serial subproblems, each to be solved in a time slot t [4].

$$\text{maximize:} \quad \sum_{j=1}^{K} \mathbb{E}\{\log(W_j)|\boldsymbol{\xi}_{[t-1]}\} \tag{4.8}$$

subject to:

$$\begin{aligned} W_j &= W_j^- + \xi_{0,j}\rho_{0,j}R_{0,j} + \xi_{1,j}\rho_{1,j}GR_{1,j}, j = 1, \ldots, K \\ &\sum_{j=1}^{K} \rho_{i,j} \le 1, \quad i = 0, 1 \\ &\rho_{i,j} \ge 0, \; i = 0, 1, \; j = 1, \ldots, K, \end{aligned}$$

where $\mathbb{E}\{\log(W_j)|\boldsymbol{\xi}_{[t-1]}\}$ denotes the *conditional expectation* of $\log(W_j)$ given realization $\boldsymbol{\xi}_{[t-1]}$. W_j^- is known given the realization. When $t = 1$, the conditional expectation becomes an unconditional expectation.

Since a CR user has only one transceiver, it can operate on either one or more licensed channels (i.e., connecting to the FBS) or the common channel (i.e., connecting to the MBS), but not both simultaneously. Assume CR user j operates on the common channel with probability p_j and one or more licensed channels with probability q_j. We then rewrite problem (4.8) as

$$\text{maximize:} \quad \sum_{j=1}^{K} [p_j \bar{P}_{0,j} \log(W_j^- + \rho_{0,j}R_{0,j}) + q_j \bar{P}_{1,j} \log(W_j^- + \rho_{1,j}GR_{1,j})] \tag{4.9}$$

subject to:

$$\begin{aligned} &\sum_{j=1}^{K} \rho_{i,j} \le 1, \quad i = 0, 1 \\ &p_j + q_j = 1, \quad j = 1, \ldots, K \\ &\rho_{i,j}, \; p_j, \; q_j \ge 0, \quad i = 0, 1, \; j = 1, \ldots, K. \end{aligned}$$

Properties

In this section, we analyze the formulated problem (4.9) and derive its properties. We have Lemmas 4.1, 4.2, and 4.3 and Theorem 4.1 and provide the proofs in the following.

Lemma 4.1. *Problem* (4.9) is a convex optimization problem.

Proof. First, it can be shown that the single term

$$p_j \bar{P}_{0,j} \log(W_j^- + \rho_{0,j}R_{0,j}) + q_j \bar{P}_{1,j} \log(W_j^- + \rho_{1,j}GR_{1,j})$$

is a concave function, because its *Hessian matrix* is negative semi-definite. Then, the objective function is concave since the sum of concave functions is also concave. Finally, all the constraints are linear. We conclude that problem (4.9) is convex with a unique optimal solution. □

Lemma 4.2. *If* $[\rho, p, q]$ *is a feasible solution to problem* (4.9), then $[\rho, q, p]$ is also feasible.

Proof. Since $[\rho, p, q]$ is feasible, we have $p + q = 1$. Switching the two probabilities, we still have $q + p = 1$. Therefore, the derived new solution is also feasible. □

Lemma 4.3. *Let the optimal solution be* $[\rho^*, p^*, q^*]$. *If* $p_j^* \geq q_j^*$, *then the term* $\bar{P}_{0,j} \log(W_j^- + \rho_{0,j}^* R_{0,j})$ *is greater than or equal to* $\bar{P}_{1,j} \log(W_j^- + \rho_{1,j}^* G R_{1,j})$. *And vice versa.*

Proof. Assume $\bar{P}_{0,j} \log(W_j^- + \rho_{0,j}^* R_{0,j})$ is less than $\bar{P}_{1,j} \log(W_j^- + \rho_{1,j}^* G R_{1,j})$. Since $p_j^* \geq q_j^*$, the sum of the products

$$p_j^* \bar{P}_{0,j} \log(W_j^- + \rho_{0,j}^* R_{0,j}) + q_j^* \bar{P}_{1,j} \log(W_j^- + \rho_{1,j}^* G R_{1,j})$$

is smaller than the sum of the products

$$q_j^* \bar{P}_{0,j} \log(W_j^- + \rho_{0,j}^* R_{0,j}) + p_j^* \bar{P}_{1,j} \log(W_j^- + \rho_{1,j}^* G R_{1,j}).$$

Thus we can obtain an objective value larger than the optimum by switching the values of p_j^* and q_j^*, which is still feasible according to Lemma 4.2. This conflicts with the assumption that $[\rho^*, p^*, q^*]$ is optimal. The reverse statement can be proved similarly. □

Theorem 4.1. *Let the optimal solution be* $[\rho^*, p^*, q^*]$. *If* $p_j^* > q_j^*$, *then we have* $p_j^* = 1$ *and* $q_j^* = 0$. *Otherwise, we have* $p_j^* = 0$ *and* $q_j^* = 1$.

Proof. If $p_j^* > q_j^*$, we have $\bar{P}_{0,j} \log(W_j^- + \rho_{0,j}^* R_{0,j}) \geq \bar{P}_{1,j} \log(W_j^- + \rho_{1,j}^* G R_{1,j})$ according to Lemma 4.3. Since the objective function is linear with respect to p_j and q_j, the optimal value can be achieved by setting p_j to its maximum value 1 and q_j to its minimum value 0. The reverse statement can be proved similarly. □

According to Theorem 4.1, a CR user is connected to either the MBS or the FBS for the *entire* duration of a time slot in the optimal solution. That is, it does not switch between base stations during a time slot under optimal scheduling.

Distributed Solution Algorithm

To solve problem (4.9), we define nonnegative *dual variables* $\lambda = [\lambda_0, \lambda_1]$ for the two inequality constraints. The *Lagrangian function* is

$$\begin{aligned}
&\mathcal{L}(p, \rho, \lambda) \\
&= \sum_{j=1}^{K} [p_j \bar{P}_{0,j} \log(W_j^- + \rho_{0,j} R_{0,j}) + (1 - p_j) \bar{P}_{1,j} \log(W_j^- + \rho_{1,j} G R_{1,j})] \\
&+ \lambda_0 \left(1 - \sum_{j=1}^{K} \rho_{0,j}\right) + \lambda_1 \left(1 - \sum_{j=1}^{K} \rho_{1,j}\right) \\
&= \sum_{j=1}^{K} \mathcal{L}_j(p_j, \rho_{0,j}, \rho_{1,j}, \lambda_0, \lambda_1) + \lambda_0 + \lambda_1,
\end{aligned} \tag{4.10}$$

where

$$\begin{aligned}
\mathcal{L}_j(p_j, \rho_{0,j}, \rho_{1,j}, \lambda_0, \lambda_1) = p_j \bar{P}_{0,j} &\log(W_j^- + \rho_{0,j} R_{0,j}) \\
&+ (1 - p_j) \bar{P}_{1,j} \log(W_j^- + \rho_{1,j} G R_{1,j}) \\
&- \lambda_0 \rho_{0,j} - \lambda_1 \rho_{1,j}.
\end{aligned}$$

The corresponding problem can be decomposed into K subproblems and solved iteratively. In every Step $\tau \geq 1$, for given $\lambda_0(\tau)$ and $\lambda_1(\tau)$ values, each CR user j solves the following subproblem using local information.

$$[p_j^*(\tau), \rho_{0,j}^*(\tau), \rho_{1,j}^*(\tau)] = \underset{p_j, \rho_{0,j}, \rho_{1,j} \geq 0}{\arg\max} \; \mathcal{L}_j(p_j, \rho_{0,j}, \rho_{1,j}, \lambda_0(\tau), \lambda_1(\tau)). \tag{4.11}$$

There is a unique optimal solution since the objective function in (4.11) is concave. The CR users then exchange their solutions. The *master dual problem*, for given $p(\tau)$ and $\rho(\tau)$, is:

$$\begin{aligned}
&\min_{\lambda \geq 0} \mathcal{L}(p(\tau), \rho(\tau), \lambda) \\
&= \sum_{j=1}^{K} \mathcal{L}_j(p_j(\tau), \rho_{0,j}(\tau), \rho_{1,j}(\tau), \lambda_0, \lambda_1) + \lambda_0 + \lambda_1.
\end{aligned} \tag{4.12}$$

Since the Lagrangian function is differentiable, the *gradient iteration* approach can be used.

$$\lambda_i(\tau+1) = \left[\lambda_i(\tau) - s \times \left(1 - \sum_{j=1}^{K} \rho^*_{i,j}(\tau)\right)\right]^+, \quad i = 0, 1, \tag{4.13}$$

where s is a sufficiently small positive *step size* and $[\cdot]^+$ denotes the projection onto the nonnegative axis. The updated $\lambda_i(\tau+1)$ will again be used to solve the subproblems, and so forth. Since the problem is convex, we have *strong duality*; the *duality gap* between the primal and dual problems is zero. The dual variables $\lambda(\tau)$ will converge to the optimal values as τ goes to infinity. Since the optimal solution to (4.11) is unique, the primal variables $p(\tau)$ and $\rho_{i,j}(\tau)$ will also converge to their optimal values when τ is sufficiently large.

The distributed solution procedure is presented in Algorithm 8. In the table, Steps 3–9 solve the subproblem in (4.11); Step 10 updates the dual variables. The threshold ϕ is a prescribed small value with $0 \le \phi \ll 1$. The algorithm terminates when the dual variables are sufficiently close to the optimal values.

Algorithm 1: Algorithm for the Case of Single FBS ([3], © 2012 IEEE)

1 Set $\tau = 0$, $\lambda_0(0)$ and $\lambda_1(0)$ to some nonnegative value ;
2 **repeat**
3 $\quad \rho_{0,j}(\tau) = \left[\frac{\bar{P}_{0,j}}{\lambda_0(\tau)} - \frac{W_j^-}{R_{0,j}}\right]^+, \rho_{1,j}(\tau) = \left[\frac{\bar{P}_{1,j}}{\lambda_1(\tau)} - \frac{W_j^-}{R_{1,j}G}\right]^+$;
4 $\quad$ **if** $\left(\left[\bar{P}_{0,j}\log(W_j^- + \rho_{0,j}(\tau)R_{0,j}) - \lambda_0(\tau)\rho_{0,j}(\tau)\right] >\right.$
5 $\quad\quad \left.\left[\bar{P}_{1,j}\log(W_j^- + \rho_{1,j}(\tau)GR_{1,j}) - \lambda_1(\tau)\rho_{1,j}(\tau)\right]\right)$ **then**
6 $\quad\quad$ Set $p_j(\tau) = 1$ and $\rho_{1,j}(\tau) = 0$;
7 $\quad$ **else**
8 $\quad\quad$ Set $p_j(\tau) = 0$ and $\rho_{0,j}(\tau) = 0$;
9 $\quad$ **end**
10 $\quad$ MBS updates $\lambda_i(\tau+1)$ as in (4.13) ;
11 $\quad \tau = \tau + 1$;
12 **until** $\left(\sum_{i=0}^{1}(\lambda_i(\tau+1) - \lambda_i(\tau))^2 > \phi\right)$;

4.2.2 *Case of Multiple Noninterfering FBS's*

We next consider the case of $N > 1$ noninterfering FBS's. The coverages of the FBS's do not overlap with each other, as FBS 1 and 2 in Fig. 4.1. Consequently, each FBS can use all the available licensed channels without interfering other FBS's. Assume each CR user knows the nearest FBS and is associate with it. Let $\mathcal{U}_i$ denote the set of CR users associated with FBS i. The resource allocation problem becomes:

$$\text{maximize:} \quad \sum_{j=1}^{K} p_j \bar{P}_{0,j} \log(W_j^- + \rho_{0,j} R_{0,j}) \tag{4.14}$$

$$+ \sum_{i=1}^{N} \sum_{j \in \mathcal{U}_i} q_j \bar{P}_{i,j} \log(W_j^- + \rho_{i,j} G R_{i,j})$$

subject to:

$$\sum_{j=1}^{K} \rho_{0,j} \le 1$$

$$\sum_{j \in \mathcal{U}_i} \rho_{i,j} \le 1, \quad i = 1, \ldots, N$$

$$p_j + q_j = 1, \quad j = 1, \ldots, K$$

$$\rho_{i,j}, \; p_j, \; q_j \ge 0, \quad i = 1, \ldots, N, \; j = 1, \ldots, K.$$

Since all the available channels can be allocated to each FBS with spatial reuse, problem (4.14) can be solved using the algorithm in Algorithm 8 with some modified notation: $\rho_{1,j}(\tau)$ now becomes $\rho_{i,j}(\tau)$ and $\lambda_1(\tau)$ becomes $\lambda_i(\tau)$, $i = 1, \ldots, N$. The dual variables are iteratively updated as:

$$\lambda_0(\tau + 1) = \left[\lambda_0(\tau) - s \times \left(1 - \sum_{j=1}^{K} \rho_{0,j}^*(\tau) \right) \right]^+ \tag{4.15}$$

$$\lambda_i(\tau + 1) = \left[\lambda_i(\tau) - s \times \left(1 - \sum_{j \in \mathcal{U}_i} \rho_{i,j}^*(\tau) \right) \right]^+, \quad i = 1, \ldots, N. \tag{4.16}$$

The modified solution algorithm is presented in Algorithm 9. As in the case of single FBS, the algorithm is jointly executed by the CR users and MBS, by iteratively updating the dual variables $\lambda_0(\tau)$ and $\lambda_i(\tau)$'s, and the resource allocations $\rho_{0,j}^*(\tau)$ and $\rho_{i,j}^*(\tau)$'s. It can be shown that the distributed algorithm can produce the optimal solution for problem (4.14).

4.2.3 Case of Multiple Interfering FBS's

Formulation

Finally, we consider the case of multiple interfering FBS's. Assume that the coverages of some FBS's overlap with each other, as FBS 3 and 4 in Fig. 4.1. They cannot use the same channel simultaneously, but have to compete for the available channels in the transmission phase. Define *channel allocation variables* $c_{i,m}$ for time slot t as:

Algorithm 2: Algorithm for the Case of Multiple NonInterfering FBS's ([3], © 2012 IEEE)

Set $\tau = 0$, and $\lambda_0(0)$ and $\lambda_i(0)$ to some nonnegative values, for all i ;
repeat
  $\rho_{0,j}(\tau) = \left[\frac{\bar{P}_{0,j}}{\lambda_0(\tau)} - \frac{W_j^-}{R_{0,j}}\right]^+$, $\rho_{i,j}(\tau) = \left[\frac{\bar{P}_{i,j}}{\lambda_i(\tau)} - \frac{W_j^-}{R_{i,j}G}\right]^+$;
  if $\left(\left[\bar{P}_{0,j}\log(W_j^- + \rho_{0,j}(\tau)R_{0,j}) - \lambda_0(\tau)\rho_{0,j}(\tau)\right] >\right.$
   $\left.\left[\bar{P}_{i,j}\log(W_j^- + \rho_{i,j}(\tau)GR_{i,j}) - \lambda_i(\tau)\rho_{i,j}(\tau)\right]\right)$ **then**
    Set $p_j(\tau) = 1$ and $\rho_{i,j}(\tau) = 0$;
  else
    Set $p_j(\tau) = 0$ and $\rho_{0,j}(\tau) = 0$;
  end
  MBS updates $\lambda_i(\tau + 1)$ as in (4.15) and (4.16) ;
  $\tau = \tau + 1$;
until $\left(\sum_{i=0}^{N}(\lambda_i(\tau+1) - \lambda_i(\tau))^2 > \phi\right)$;

FBS 1 FBS 2 FBS 3 FBS 4

Fig. 4.3 Interference graph for the femtocell CR network shown in Fig. 4.1 ([3], © 2012 IEEE)

$$c_{i,m} = \begin{cases} 1, & \text{if channel } m \text{ is allocated to FBS } i \\ 0, & \text{otherwise.} \end{cases} \tag{4.17}$$

Given an allocation, the expected number of available channels for FBS i is $G_i = \sum_{m \in \mathcal{A}(t)} c_{i,m} P_m^A$.

We use *interference graph* to model the case of overlapping coverages, which is defined below.

Definition 4.1. An *interference graph* $G_I = (V_I, E_I)$ is an undirected graph where each vertex represents an FBS and each edge indicates interference between the two end FBS's.

For example given in Fig. 4.1, we can derive an interference graph as shown in Fig. 4.3. FBS 3 and 4 cannot use the same channel simultaneously, as summarized in the following lemma.

Lemma 4.4. *If channel m is allocated to FBS i, the neighboring vertices of FBS i in the interference graph G_I, denoted as $\mathcal{R}(i)$, cannot use the same channel m simultaneously.*

Further define index variables d_i^k as:

$$d_i^k = \begin{cases} 1, & \text{if FBS } i \text{ is an endpoint of link } k \in G_I \\ 0, & \text{otherwise.} \end{cases} \tag{4.18}$$

The interference constraint can be described as:

$$\sum_{i=1}^{N} d_i^k c_{i,m} \leq 1, \text{ for } m = 0, 1, \ldots, M, \text{ and } k \in G_I. \tag{4.19}$$

We then have the following problem formulation.

$$\begin{aligned} \text{maximize: } & \sum_{j=1}^{K} p_j \bar{P}_{0,j} \log(W_j^- + \rho_{0,j} R_{0,j}) + \\ & \sum_{i=1}^{N} \sum_{j \in \mathcal{U}_i} q_j \bar{P}_{i,j} \log(W_j^- + \rho_{i,j} G_i R_{i,j}) \end{aligned} \tag{4.20}$$

subject to:

$$\begin{aligned} & \sum_{j=1}^{K} \rho_{0,j} \leq 1 \\ & \sum_{j \in \mathcal{U}_i} \rho_{i,j} \leq 1, \quad i = 1, \ldots, N \\ & p_j + q_j = 1, \quad j = 1, \ldots, K \\ & G_i = \sum_{m \in \mathcal{A}(t)} c_{i,m} P_m^A, \quad i = 1, \ldots, N \\ & \sum_{i=1}^{N} d_i^k c_{i,m} \leq 1, m = 0, \ldots, M, \text{ for link } k \in G_I, \\ & \rho_{i,j}, p_j, q_j, c_{i,m} \geq 0, \quad i = 1, \ldots, N, \; j = 1, \ldots, K, \; m = 0, \ldots, M. \end{aligned}$$

Algorithm 3: Channel Allocation Algorithm for Case of Interfering FBS's ([3], © 2012 IEEE)

Initialize $\boldsymbol{c}$ to a zero matrix, FBS set $\mathcal{N} = \{1, \ldots, N\}$, and FBS-channel set $\mathcal{C} = \mathcal{N} \times \mathcal{A}(t)$;
while ($\mathcal{C}$ *is not empty*) **do**
  Find FBS-channel pair $\{i', m'\}$, such that $\{i', m'\} = \arg\max_{\{i,m\} \in \mathcal{C}} \{Q(\boldsymbol{c} + \boldsymbol{e}_{i,m}) - Q(\boldsymbol{c})\}$;
  Set $\boldsymbol{c} = \boldsymbol{c} + \boldsymbol{e}_{i',m'}$;
  Remove $\{i', m'\}$ from $\mathcal{C}$;
  Remove $\mathcal{R}(i') \times m'$ from $\mathcal{C}$;
end

Solution Algorithm

The optimal solution to problem (4.20) depends on the channel allocation variables $c_{i,m}$. Problem (4.20) can be solved with the algorithm in Algorithm 9 if the $c_{i,m}$'s are known. Let $Q(\boldsymbol{c})$ be the suboptimal objective value for a given channel allocation $\boldsymbol{c}$, where $\boldsymbol{c} = [\vec{c}_1, \vec{c}_2, \ldots, \vec{c}_N]$ and $\vec{c}_i$ is a vector of elements $c_{i,m}$, for FBS i and channels $m \in \mathcal{A}(t)$. If all the FBS's are disjointedly distributed with no overlap, each FBS can use all the available channels. We have $c_{i,m} = 1$ for all i and $m \in \mathcal{A}(t)$, i.e., it is reduced to the case in Sect. 4.2.2.

To solve problem (4.20), we first apply a *greedy algorithm* to allocate the available channels in $\mathcal{A}(t)$ to the FBS's (i.e., to determine $\boldsymbol{c}$). We then apply the algorithm in Algorithm 9 with the computed $\boldsymbol{c}$ to obtain a near-optimal solution. Let $\boldsymbol{e}_{i,m}$ be a matrix with 1 at position $\{i, m\}$ and 0 at all other positions, representing the allocation of channel $m \in \mathcal{A}(t)$ to FBS i. The greedy channel allocation algorithm is given in Algorithm 10, where the FBS-channel pair that can achieve the largest increase in $Q(\cdot)$ is chosen in each iteration. The worst case complexity of the greedy algorithm is $O(N^2M^2)$.

Performance Lower Bound

We next present a lower bound for the greedy algorithm. Let $e(l)$ be the lth FBS-channel pair chosen in the greedy algorithm, and π_l denote the sequence $\{e(1), e(2), \ldots, e(l)\}$. The increase in object value (4.20) due to the lth allocated FBS-channel pair is denoted as:

$$\Delta_l := \Delta(\pi_l, \pi_{l-1}) = Q(\pi_l) - Q(\pi_{l-1}). \tag{4.21}$$

Since $Q(\pi_0) = Q(\emptyset) = 0$, we have

$$\begin{aligned}
\sum_{l=1}^{L} \Delta_l &= Q(\pi_L) - Q(\pi_{L-1}) + \cdots + Q(\pi_1) - Q(\pi_0) \\
&= Q(\pi_L) - Q(\pi_0) \\
&= Q(\pi_L).
\end{aligned}$$

For two FBS-channel pairs $e(l)$ and $e(l')$, we say $e(l)$ *conflicts with* $e(l')$ when there is an edge connecting the FBS in $e(l)$ and the FBS in $e(l')$ in the interference graph G_I, and the two FBS's choose the same channel. Let Ω be the global optimal solution. We define ω_l as the subset of Ω that conflicts with allocation $e(l)$ but not with the previous allocations $\{e(1), e(2), \ldots, e(l-1)\}$.

Lemma 4.5. *Assume the greedy algorithm in Algorithm 10 stops in L steps. The global optimal solution Ω can be partitioned into L non-overlapping subsets ω_l, $l = 1, 2, \ldots, L$.*

Proof. According to the definition of ω_l, the L subsets of the optimal solution Ω do not intersect with each other. Assume the statement is false, then the union of these L subsets is not equal to the optimal set Ω. Let the *set difference* be $\omega_{L+1} = \Omega \setminus (\cup_{l=1}^{L}\omega_l)$. By definition, ω_{L+1} does not conflict with the existing L allocations $\{e(1), \ldots, e(L)\}$, meaning that the greedy algorithm can continue to at least the $(L + 1)$th step. This conflicts with the assumption that the greedy algorithm stops in L steps. It follows that $\Omega = \cup_{l=1}^{L}\omega_l$. □

Let $\Delta(\pi_2, \pi_1) = Q(\pi_2) - Q(\pi_1)$ denote the difference between two feasible allocations π_1 and π_2. We next derive a lower bound on the performance of the greedy algorithm. We assume two properties for function $\Delta(\pi_2, \pi_1)$ in the following.

Property 4.1. *Consider FBS-channel pair sets π_1, π_2, and σ, satisfying $\pi_1 \subseteq \pi_2$ and $\sigma \cap \pi_2 = \emptyset$. We have $\Delta(\pi_2 \cup \sigma, \pi_1 \cup \sigma) \leq \Delta(\pi_2, \pi_1)$.*

Property 4.2. *Consider FBS-channel pair sets π, σ_1, and σ_2 satisfying $\sigma_1 \cap \pi = \emptyset$, $\sigma_2 \cap \pi = \emptyset$, and $\sigma_1 \cap \sigma_2 = \emptyset$. We have $\Delta(\sigma_1 \cup \sigma_2 \cup \pi, \pi) \leq \Delta(\sigma_1 \cup \pi, \pi) + \Delta(\sigma_2 \cup \pi, \pi)$.*

In Property 4.1, we have $\sigma \cap \pi_1 = \emptyset$ since $\pi_1 \subseteq \pi_2$ and $\sigma \cap \pi_2 = \emptyset$. This property states that the incremental objective value does not get larger as more channels are allocated and as the objective value gets larger. Property 4.2 states that the incremental objective value achieved by allocating multiple FBS-channel pair sets does not exceed the sum of the incremental objective values achieved by allocating each individual FBS-channel pair set. These are generally true for many resource allocation problems [25].

In the greedy algorithm, since we choose the maximum incremental allocation at every step, we have Lemma 4.6 that directly follows Step 3 in Algorithm 10.

Lemma 4.6. *For any FBS-channel pair $\sigma \in \omega_l$, we have $Q(\pi_{l-1} \cup \sigma) - Q(\pi_{l-1}) = \Delta(\pi_{l-1} \cup \sigma, \pi_{l-1}) \leq \Delta_l$.*

Lemma 4.7. *Assume the greedy algorithm stops in L steps, we have*

$$Q(\Omega) \leq Q(\pi_L) + \sum_{l=1}^{L} \sum_{\sigma \in \omega_l} \Delta(\sigma \cup \pi_{l-1}, \pi_{l-1}).$$

Proof. The following inequalities hold true according to the properties of the $\Delta(\cdot, \cdot)$ function:

$$\begin{aligned}
&Q((\cup_{i=l+1}^{L}\omega_i) \cup \pi_l) \\
&= Q((\cup_{i=l+2}^{L}\omega_i) \cup \pi_l) + \Delta((\cup_{i=l+1}^{L}\omega_i) \cup \pi_l, (\cup_{i=l+2}^{L}\omega_i) \cup \pi_l) \\
&\leq Q((\cup_{i=l+2}^{L}\omega_i) \cup \pi_l) + \Delta(\omega_{l+1} \cup \pi_l, \pi_l) \\
&\leq Q((\cup_{i=l+2}^{L}\omega_i) \cup \pi_{l+1}) + \Delta(\omega_{l+1} \cup \pi_l, \pi_l) \\
&\leq Q((\cup_{i=l+2}^{L}\omega_i) \cup \pi_{l+1}) + \sum_{\sigma \in \omega_{l+1}} \Delta(\sigma \cup \pi_l, \pi_l).
\end{aligned}$$

We have $\pi_0 = \emptyset$ and $\omega_{L+1} = \emptyset$ (see Lemma 4.5). With induction from $l = 0$ to $l = L-1$, we have $Q((\cup_{i=1}^{L}\omega_i)\cup\emptyset) = Q(\Omega)$ and $Q(\Omega) \leq Q(\pi_L)+\sum_{l=1}^{L}\sum_{\sigma\in\omega_l}\Delta(\sigma\cup \pi_{l-1}, \pi_{l-1})$. □

Lemma 4.8. *The maximum size of ω_l is equal to the degree, in the interference graph G_I, of the FBS selected in the lth step of the greedy algorithm, which is denoted as $D(l)$.*

Proof. Once FBS i is allocated with channel m, the neighboring FBS's in G_I, $\mathcal{R}(i)$, cannot use the same channel m anymore due to the interference constraint. The maximum number of FBS-channel pairs that conflict with the selected FBS-channel pair $\{i, m\}$, i.e., the maximum size of ω_l, is equal to the degree of FBS i in G_I. □

Then we have Theorem 4.2 that provides a lower bound on the objective value achieved by the greedy algorithm given in Algorithm 10.

Theorem 4.2. *The greedy algorithm can achieve an objective value that is at least $\frac{1}{1+D_{\max}}$ of the global optimum, where $D_{\max}$ is the maximum node degree in the interference graph G_I of the femtocell CR network.*

Proof. According to Lemmas 4.7 and 4.8, we have:

$$\begin{aligned} Q(\Omega) &\leq Q(\pi_L) + \sum_{l=1}^{L} D(l)\Delta_l \\ &= Q(\pi_L) + \bar{D}\sum_{l=1}^{L}\Delta_l \\ &= (1+\bar{D})Q(\pi_L), \end{aligned} \tag{4.22}$$

where $\bar{D} = \sum_{l=1}^{L} D(l)\Delta_l / \sum_{l=1}^{L}\Delta_l$. The second equality is due to the facts that $\sum_{l=1}^{L}\Delta_l = Q(\pi_L)$.

To further simplify the bound, we replace $D(l)$ with the maximum node degree $D_{\max}$. We then have

$$\bar{D} \leq \frac{\sum_{l=1}^{L} D_{\max}\Delta_l}{\sum_{l=1}^{L}\Delta_l} = D_{\max}$$

and

$$\frac{1}{1+D_{\max}}Q(\Omega) \leq Q(\pi_L) \leq Q(\Omega), \tag{4.23}$$

which provides a lower bound on the performance of the greedy algorithm. □

When there is a single FBS in the CR network, we have $D_{\max} = 0$ and $Q(\pi_L) = Q(\Omega)$ according to Theorem 4.2. The proposed algorithm produces the optimal solution. In the case of multiple noninterfering FBS's, we still have $D_{\max} = 0$ and can obtain the optimal solution using the proposed algorithm. For the femtocell

CR network given in Fig. 4.1 (with interference graph shown in Fig. 4.3), we have $D_{\max} = 1$ and the low bound is a half of the global optimal. Note that (4.22) provides a tighter bound for the optimum than (4.23), but with higher complexity. These are interesting performance bounds since they bound the achievable video quality, an application layer performance measure, rather than lower layer metrics (e.g., bandwidth or time share).

4.3 Simulation Results

We evaluate the performance of the proposed algorithms using MATLAB and JSVM 9.13 Video codec. Two scenarios are used in the simulations: a single FBS CR network and a CR network with interfering FBS's. In every simulation, we compare the proposed algorithms with the following three more straightforward heuristic schemes:

- Heuristic 1 based on *equal allocation*: each CR user chooses the better channel (i.e., the common channel or a licensed channel) based on the channel conditions; time slots are equally allocated among active CR users;
- Heuristic 2 exploiting *multiuser diversity*: the MBS and each FBS chooses one active CR user with the best channel condition; the entire time slot is allocated to the selected CR user.
- *SCA-MAC* proposed in [66]: with this scheme, the successful transmission rate is evaluated based on channel packet loss rate and collision probability with primary users; the channel-user pair with the highest transmission probability is selected.

We choose SCA-MAC because it adopts similar models and assumptions as in this chapter. Once of the channels are selected, the same distributed algorithm is used for scheduling video data for all the three schemes.

We adopt the Raleigh block fading model and the packet loss probability is between [0.004 and 0.028]. The frame rate is set to 30 fps and the GoP size is 16. The base layer mode is set to be AVC compatible. The motion search mode is set to Fast Search with search range 32. Each point in the figures presented in this section is the average of 10 simulation runs with different random seeds. We plot 95 % confidence intervals in the figures, which are generally negligible.

4.3.1 Case of Single FBS

In the first scenario, there are $M = 8$ channels and the channel parameters P_{01}^m and P_{10}^m are set to 0.4 and 0.3, respectively, for all m. The maximum allowable collision probability γ_m is set to 0.2 for all m. There is one FBS and three active CR users. Three Common Intermediate Format (CIF, 352×288) video sequences are streamed to the CR users, i.e., *Bus* to CR user 1, *Mobile* to CR user 2, and *Harbor* to CR

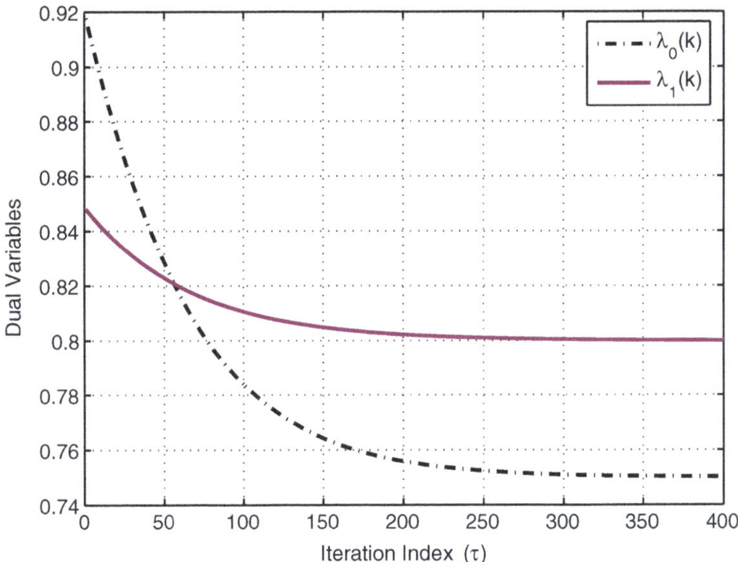

Fig. 4.4 Convergence of the two dual variables in the single FBS case ([3], © 2012 IEEE)

user 3. We have $T = 10$ as the delivery deadline. Both probabilities of false alarm ϵ and miss detection δ are set to 0.3 for all the FBS's and CR users, unless otherwise specified.

First we investigate the convergence of the distributed algorithm. The traces of the two dual variables are plotted in Fig. 4.4. To improve the convergence speed, the correlation in adjacent time slots can be exploited. In particular, we set the optimal values for the optimization variables in the previous time slot as the initialization values for the variables in the current time slot. By doing so, the convergence speed can be improved. It can be seen that both dual variables converge to their optimal values after 300 iterations. After convergence, the optimal solution for the primary problem can be obtained.

Our proposed scheme achieves the best performance among the three algorithms, with up to 4.3 dB improvement over the two heuristic schemes and up to 2.5 dB over SCA-MAC. Such gains are significant with regard to video quality, since a 0.5 dB difference is distinguishable by human eyes. Compared to the two heuristic schemes and SCA-MAC, the video quality of our proposed scheme is well balanced among the three users, indicating better fairness performance.

In Fig. 4.5, we examine the impact of the number of channels M on received video quality. First, we validate the video quality measure used in our formulation by comparing the PSNR value computed using (4.5) with that computed from real decoded video frames. The average PSNR for three received videos are plotted in the figure. It can be seen that the real PSNRs are very close to those predicted by (4.5), with overlapping confidence intervals. This is also consistent with the results shown

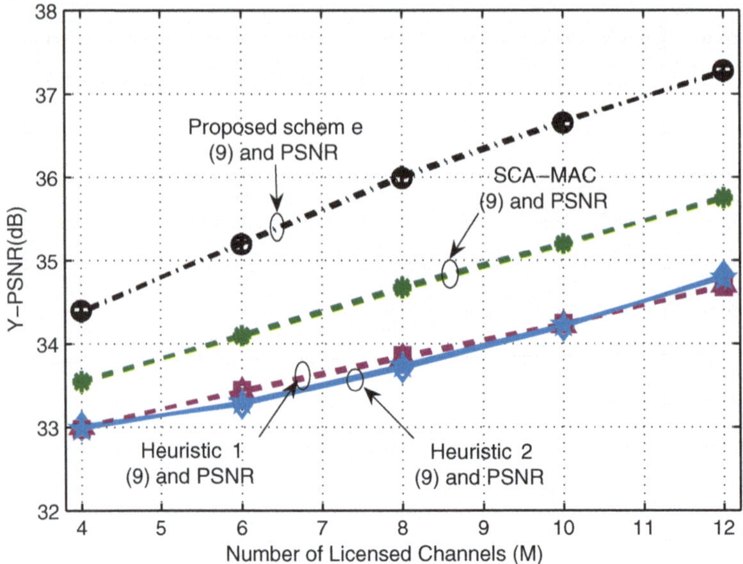

Fig. 4.5 Single FBS: received video quality vs. number of channels (computed with (9) and measured by PSNR) ([3], © 2012 IEEE)

in Fig. 4.2. Second, as expected, the more licensed channels, the more spectrum opportunities for CR users, and the higher PSNR for received videos. SCA-MAC performs better than two heuristics, but is inferior to the proposed scheme.

We also plot the MS-SSIM of the received videos at the three CR users in Fig. 4.6 [65]. Similar observations can be made from the MS-SSIM plot. All MS-SSIMs for the four curves are more than 0.97 and very close to 1. The proposed scheme still outperforms the other three schemes. In the remaining figures, we will use model predicted PSNR values, since the model (4.5) is sufficient to predict the real video quality.

In Fig. 4.7, we demonstrate the impact of channel utilization η on received video quality. The average PSNRs achieved by the four schemes are plotted when η is increased from 0.3 to 0.7. Intuitively, a smaller η allows more spectrum opportunities for video transmission. This is shown in the figure where all the three curves decrease as η gets larger. The performance of both heuristics are close and the proposed scheme achieves a gain about 3 dB over the heuristics and 2 dB over SCA-MAC.

We also compare the MGS and FGS videos while keeping other parameters identical. We find that MGS video achieves over 0.5 dB gain in video quality over FGS video. The results are omitted for brevity.

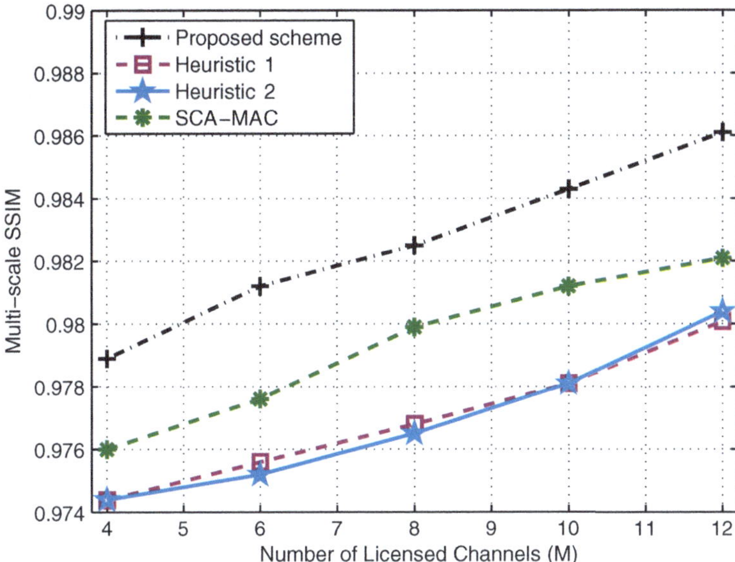

Fig. 4.6 Single FBS: received video quality vs. number of channels (measured by MS-SSIM) ([3], © 2012 IEEE)

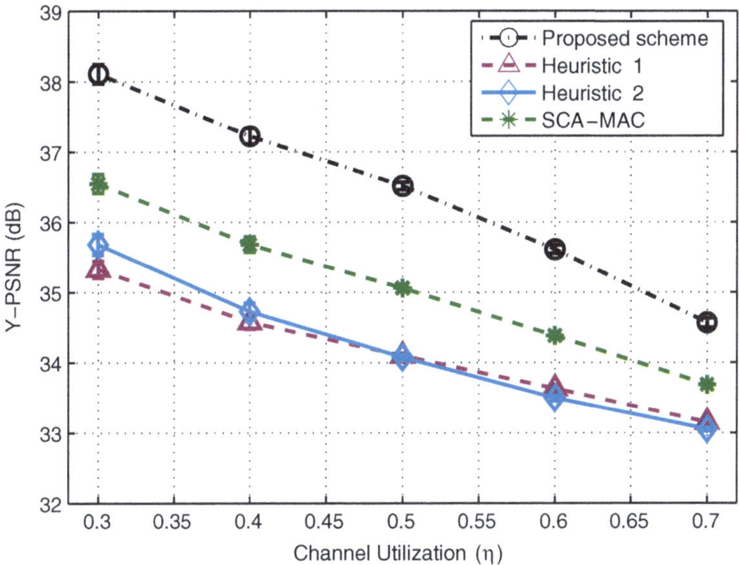

Fig. 4.7 Single FBS: received video quality versus channel utilization ([3], © 2012 IEEE)

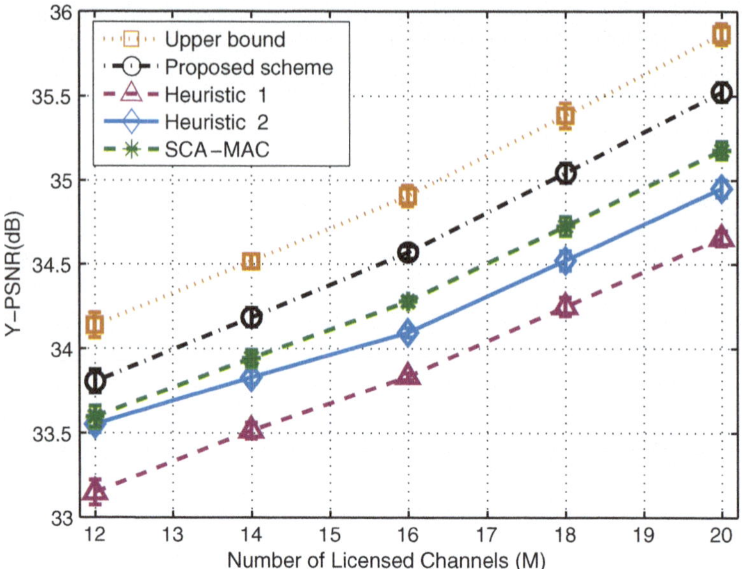

Fig. 4.8 Interfering FBS's: received video quality versus number of channels ([3], © 2012 IEEE)

4.3.2 Case of Interfering FBS's

We next investigate the second scenario with three FBS's, and each FBS has three active CR users. Each FBS streams three different videos to the corresponding CR users. The coverages of FBS 1 and 2 overlap with each other, and the coverages of FBS 2 and 3 overlap with each other.

In Fig. 4.8, we examine the impact of the number of channels M on the received video quality. The average PSNRs of all the active CR users are plotted in the figure when we increase M from 12 to 20 with step size 2. As mentioned before, more channels imply more transmission opportunities for video transmission. In this scenario, heuristic 2 (with a multiuser diversity approach) outperforms heuristic 1 (with an equal allocation approach). But its PSNRs are still about $0.3 \sim 0.5$ dB lower that those of the proposed algorithm. The proposed scheme has up to 0.4 dB improvement over SCA-MAC. In Fig. 4.8, we also plot an upper bound on the optimal objective value, which is obtained as in (4.22). It can be seen that the performance of our proposed scheme is close to optimal solution since the gap between the upper bound and our scheme is generally small (about 0.5 dB).

Next, we examine the impact of sensing errors on the received video quality. In Fig. 4.9, we test five pairs of $\{\epsilon, \delta\}$ values: $\{0.2, 0.48\}$, $\{0.24, 0.38\}$, $\{0.3, 0.3\}$, $\{0.38, 0.24\}$, and $\{0.48, 0.2\}$. It is interesting to see that the performance of all the four schemes get worse when the probability of one of the two sensing errors gets large. We can trade-off between false alarm and miss detection probabilities to find the optimal operating point for the spectrum sensors. Moreover, the dynamic range

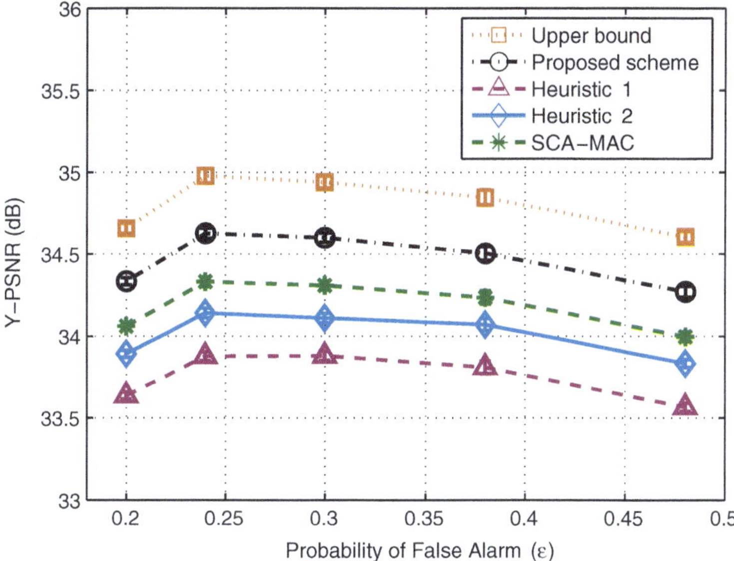

Fig. 4.9 Interfering FBS's: received video quality versus sensing error probability ([3], © 2012 IEEE)

of video quality is not big for the range of sensing errors simulated, compared to that in Fig. 4.8. This is because both sensing errors are modeled and treated in the algorithms. Again, our proposed scheme outperforms the two heuristic schemes and SCA-MAC with considerable margins for the entire range.

We also investigate the impact of the bandwidth of the common channel B_0. In this simulation, we fix B_1 at 0.3 Mbps and increase B_0 from 0.1 to 0.5 Mbps with step size 0.1 Mbps. The results are shownin Fig. 4.10. We notice that the average video quality increases rapidly as the common channel bandwidth is increased from 0.1 to 0.3 Mbps. Beyond 0.3 Mbps, the increases of the PSNR curves slow down and the curves get flat. This implies that a very large bandwidth for the common channel is not necessary, since the gain for additional bandwidth diminishes as B_0 gets large. Again, the proposed scheme outperforms the other three schemes and the gap between our scheme and the upper bound is small.

Next, we stop the distributed algorithm after a fixed amount of time, and evaluate the suboptimal solutions. In particular, we vary the duration of time slots, and let the distributed algorithm run for 5 % of the time slot duration at the beginning of the time slot. Then the solution obtained this way will be used for the video data transmissions. The results are shown in Fig. 4.11. It can be seen that when the time slot is 5 ms, the algorithm does not converge after $5\,\% \times 5 = 0.25$ ms and the PSNR produced by the distributed algorithm is close to that of Heuristic 1, and lower than those of Heuristic 2 and SCA-MAC. When the time slot is sufficiently large, the algorithm can get closer to the optimal and the proposed algorithm produces better video quality as compared

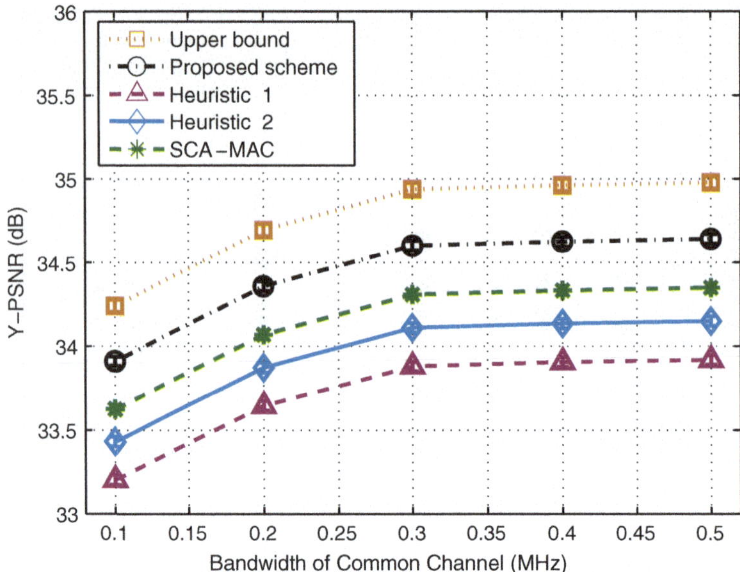

Fig. 4.10 Interfering FBS's: received video quality versus bandwidth of the common channel ([3], © 2012 IEEE)

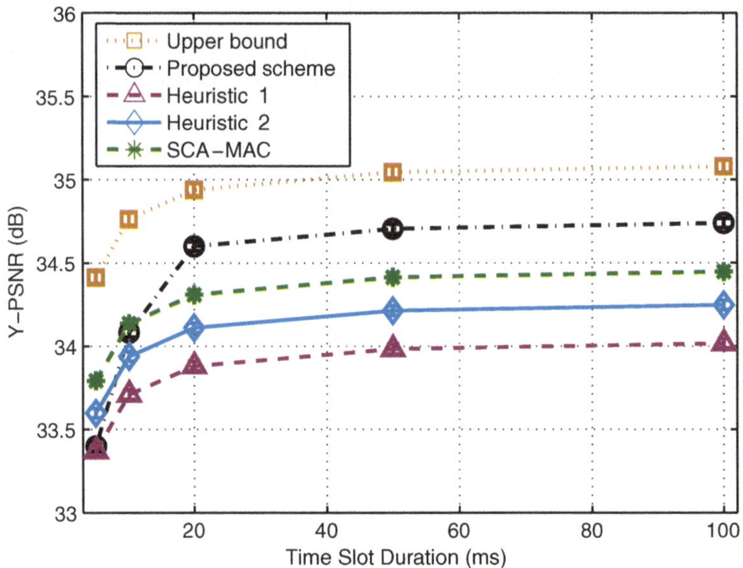

Fig. 4.11 Video quality achieved by the algorithms when they are only executed for 5 % of the time slot duration ([3], © 2012 IEEE)

to the two heuristic algorithms and SCA-MAC. Beyond 20 ms, the increase in PSNR is small since all the curves gets flat. Therefore, the proposed algorithm could be useful even when there is no time for it to fully converge to the optimal.

During the simulations, we find the collision rate with primary users are strictly kept below the prescribed collision tolerance γ. These results are omitted for brevity.

4.4 Related Work

Femtocells have received considerable interest from both industry and academia. Comprehensive overviews of technical challenges, requirements, and some preliminary solutions to femtocell networks can be found in [61, 62, 67]. Since femtocells can use the same channels as conventional cellular networks, considerable research efforts were focused on interference analysis and mitigation [68–79]. In [68], a distributed utility-based SINR adaptation scheme at femtocells was proposed to alleviate cross-tier interference at the macrocell from co-channel femtocells. In [69], the authors proposed a fractional frequency reuse scheme to mitigate inter-femtocell interference. Co-channel interference analysis and cancellation were considered in [70]. In [71], partial GSM spectrum was reused by femtocells to mitigate the negative interference impact upon the macrocell. A solution was presented for frequency allocation and power configuration within the femtocell framework. In a recent work [72], a decentralized Q-learning approach was presented for interference management. It was shown that the Q-learning scheme yields significant gains in terms of coverage speed and precision.

Deploying femtocells by underlaying the macrocell has been proved to significantly improve indoor coverage and system capacity. However, interference mitigation in a two-tier heterogeneous network is a challenging problem. In [73], the interference from macrocell and femtocells was mitigated by a spatial channel separation scheme with codeword-to-channel mapping. In [74], the rate distribution in the macrocell was improved by subband partitioning and modest gains were achieved by interference cancellation. In [75], the interference was controlled by denying the access of femtocell base stations to protect the transmission of nearby macro base station. A novel algorithmic framework was presented in [76] for dynamic interference management to deliver QoS, fairness and high system efficiency in LTE-A femtocell networks. Requiring no modification of existing macrocells, CR was shown to achieve considerable performance improvement when applied to interference mitigation [77]. In [78], the orthogonal time-frequency blocks and transmission opportunities were allocated based on a safe/victim classification.

The high potential of CRs has attracted significant interest from the wireless community [12, 38]. However, the problem of video over CR networks has been addressed only in a few papers [1, 4, 58, 59, 80, 81]. In [58], the authors presented a dynamic channel selection scheme for CR users to transmit videos over multiple channels. In [80], Ali and Yu considered transmitting a video stream over a CR link and provided a partially observable Markov decision process (POMDP) formulation and solution procedure. In [59], a distributed joint routing and spectrum sharing algorithm for video streaming applications over CR ad hoc networks was proposed and evaluated with simulations. In our prior work [1, 4], we considered scalable video streaming in an infrastructure-based CR network and a multi-hop CR network, respectively. We developed cross-layer formulation and effective algorithms for scheduling video data with proven optimality bounds.

4.5 Conclusions

In this chapter, we investigated the problem of streaming multiple MGS videos in a femtocell CR network. We formulated a multistage stochastic programming problem considering various design factors across multiple layers. A distributed algorithm was developed that can produce optimal solutions in the case of noninterfering FBS's. In the case of interfering FBS's, a greedy algorithm was developed for near-optimal solutions with a proven lower bound. The proposed algorithms were evaluated with simulations and were shown to outperform three alternative schemes with considerable gains.

Chapter 5
Video over Multi-hop CR Networks

In this chapter, we study the more challenging problem of video over multi-hop CR networks. As illustrated in Fig. 5.1, we consider an infrastructureless multi-hop CR network co-located with one or more fixed primary networks. CR users nonintrusively exploit white spaces in the licensed bands for streaming multiple videos. The objective is two-fold: to maximize the overall video quality and to achieve fairness among the concurrent video sessions, subject to bounded interference to primary users.

We adopt Fine-Granularity-Scalability (FGS) videos to accommodate heterogeneous channel availabilities and dynamic network conditions [15]. FGS video is coded into a *base layer* (BL) and an *enhancement layer* (EL). The EL can be truncated at any bit location, while all the remaining bits are still useful for decoding. This feature simplifies the design of video streaming systems. We also consider H.264/SVC medium grain scalable (MGS) videos in this chapter. MGS is shown to achieve better rate-distortion performance over MPEG-4 FGS, although MGS only has Network Abstraction Layer (NAL) unit-based granularity [57].

In order to model and guarantee end-to-end video performance, we adopt the *amplify-and-forward* relay approach for video data, which is well studied in the context of cooperative communications [40]. Specifically, each CR node is equipped with two transceivers operating on orthogonal channels. During data transmission, a relay CR node receives data from its upstream node using one transceiver on one channel, while simultaneously amplifies and forwards the received data to its downstream node using the other transceiver operating on a different channel. This is equivalent to establishing a virtual tunnel through a multi-hop multi-channel path, as illustrated in Fig. 5.2. It is also analogous to the "cut-through switching" approach for packet switching networks [82]. In addition to allowing a neat formulation of the challenging multi-hop video streaming problem, this approach also satisfies video's needs for low latency, low jitter, and high bandwidth. Its feasibility and practical considerations for multi-hop wireless networks have been addressed in [83].

The target problem is nontrivial due to the additional dimension of network dynamics (i.e., channel availability) and the additional uncertainty (i.e., spectrum

S. Mao, *Video over Cognitive Radio Networks*, DOI: 10.1007/978-1-4614-4957-7_5,

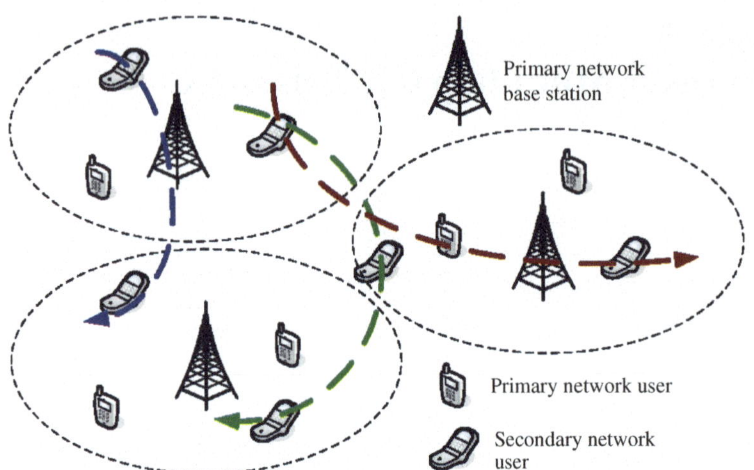

Fig. 5.1 Illustration of the multi-hop video CR network architecture ([4], ©2010 IEEE)

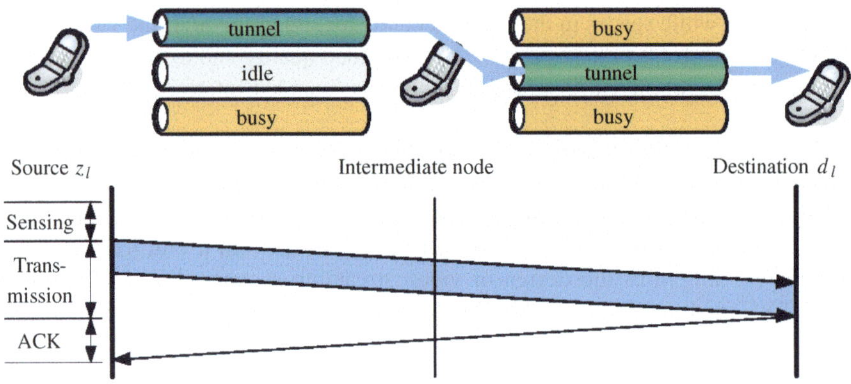

Fig. 5.2 The cut-through switching model for video data ([4], ©2010 IEEE)

sensing and sensing errors) found in CR networks. The stringent QoS requirements necessitate cross-layer optimization. The lack of centralized control also calls for distributed algorithms. We formulate streaming multiple videos over a multi-hop CR network as a *mixed integer nonlinear programming* (MINLP) problem, considering important design factors such as spectrum sensing and sensing errors, spectrum access and primary user protection, video quality and fairness, channel scheduling,, and path selection. We first develop a centralized *sequential fixing* algorithm to derive upper and lower bounds for the achievable video quality. These bounds provide useful insights into performance limits of the CR video system. We then decompose the MINLP problem into a channel scheduling problem and a path selection problem. The channel scheduling problem is solved with a greedy algorithm. For path selection, we apply *dual decomposition* and develop a distributed algorithm. We prove the optimality of the proposed approach and derive the convergence condition for the distributed algorithm. The algorithms are evaluated with extensive simulations. The distributed algorithm is shown to be highly effective for supporting concurrent video sessions in multihop CR networks, as it can achieve a performance close to that of the centralized algorithm as well as the upper bound in the cases examined.

The remainder of this chapter is organized as follows. The system model is described in Sect. 5.1. We present the problem formulation and develop the centralized algorithm in Sect. 5.2. We derive the distributed algorithm and analyze its optimality and convergence performance in Sect. 5.3. Simulation results are presented in Sect. 5.4 and related work is discussed in Sect. 5.5. Section 5.6 concludes the chapter.

5.1 System Model

5.1.1 Network Model

We consider a spectrum band consisting of M orthogonal channels with identical bandwidth [84]. As shown in Fig. 5.1, the channels are shared by K primary networks and one multi-hop CR network. The primary network base stations provide data or multimedia service to primary users. There is no fixed infrastructure in the CR network; secondary users nonintrusively explore the spectrum opportunities for unicast video communications.

Primary Networks

We assume that the M channels are allocated to K primary networks, which cover different service areas. A primary network can use any of the M channels without interfering with other primary networks. We further assume that the primary systems use a synchronous slot structure as in prior work [12, 31]. Due to primary user transmissions, the occupancy of each channel evolves following a discrete-time Markov process, as validated by recent measurement studies [12, 31, 85].

In primary network k, the status of channel m in time slot t is denoted by $S_m^k(t)$ with idle (i.e., $S_m^k(t) = 0$) and busy (i.e., $S_m^k(t) = 1$) states. Let λ_m^k and μ_m^k be the transition probability of remaining in state 0 and that from state 1 to 0, respectively, for channel m in primary network k. The utilization of channel m in primary network k, denoted by $\eta_m^k = \Pr(S_m^k = 1)$, is

$$\eta_m^k = \lim_{T\to\infty} \frac{1}{T} \sum_{t=1}^{T} S_m^k(t) = \frac{1-\lambda_m^k}{1-\lambda_m^k+\mu_m^k}. \tag{5.1}$$

The Multi-hop CR Network

Consider a multi-hop CR network co-located with the primary networks, within which $\mathcal{S}$ real-time videos are streamed among N CR nodes. Let $\mathcal{U}^k$ denote the set of CR nodes that are located within the coverage of primary network k. A video session l may be relayed by multiple CR nodes if source z_l is not a one-hop neighbor of destination d_l. We assume a *common control channel* (CCC) for the CR network [31]. We also assume the timescale of the primary channel process (or, the time slot durations) is much larger than the broadcast delays on the control channel, such that feedbacks of channel information can be received at the source nodes in a timely manner.

For CR users, each time slot consists of three phases: the spectrum sensing phase, the data transmission phase, and the acknowledgment phase. Assume that each CR user has two transceivers. In the sensing phase, one transceiver is used to sense one of the M channels, while the other is tuned to the control channel to exchange channel information with other CR users. Each video source computes the optimal path selection and channel scheduling based on sensing results. In the transmission phase, the channels assigned to a video session l at each link along the path form a virtual "tunnel" connecting source z_l and destination d_l, as illustrated in Fig. 5.2. Each node can use one or more than one channels to communicate with other nodes using a channel aggregation technique such as Orthogonal Frequency Division Multiplexing (OFDM) [31, 54]. When multiple channels are available on all the links along a path, multiple tunnels can be established and used simultaneously for a video session. In the acknowledgment phase, the destination sends ACK to the source for successfully received video packets through the same tunnel.

We adopt amplify-and-forward for video transmission [40]. During the transmission phase, one transceiver of the relay node receives video data from the upstream node on one channel, while the other transceiver of the relay node amplifies and forwards the data to the downstream node on a different, orthogonal channel. There is no need to store video packets at the relay nodes. Error detection/correction will be performed at the destination node. As a result, we can transmit through the tunnel a block of video data with minimum delay and jitter in one time slot.

Compared to the traditional "hop-centric" approach, this scheme greatly reduces the collision, contention, processing, and queueing delay induced at relay nodes [82, 83]. It is suitable for real-time data with tight delay and jitter requirements. It is especially amicable for FGS video, since a corrupted packet may still be useful for decoding. The viability, protocol-related issues, and practical considerations of this approach are addressed in [83]. The challenging issue, however, is how to set up the tunnels, while the available channels at the relays evolve over time due to primary user transmissions. We will address this issue in Sect. 5.3.

5.1.2 Spectrum Sensing

Although precise and timely channel state information is highly desirable, continuous full-spectrum sensing is hardware demanding. Without loss of generality, we assume each CR user periodically chooses one channel to sense in each time slot [18]. The index of the chosen channel by user i in time slot t is: $M_i(t) = [M_i(0) + t - 1] \bmod M$, where $M_i(0)$ is the index of the channel sensed in time slot 0.

There are two types of spectrum sensing errors: with a *false alarm*, a spectrum opportunity will be wasted, while a *miss detection* may lead to collision with primary users. Without loss of generality, we assume that the same spectrum sensing mechanism is used with identical sensing error probabilities for all CR users. Let ϵ_m and δ_m denote the probabilities of false alarm and miss detection on channel m, respectively. For time slot t, we have:

$$\begin{cases} P(W_i^m(t) = 1 | S_m^k(t) = 0) = \epsilon_m, \ m = 1, \ldots, M \\ P(W_i^m(t) = 0 | S_m^k(t) = 1) = \delta_m, \ m = 1, \ldots, M, \end{cases} \tag{5.2}$$

where $W_i^m(t)$ is user i's sensing result for channel m.

In a multi-hop CR network, the sensing results from various users may be different. Denote H_0 as the hypothesis that channel m in primary network k is idle, and H_1 the hypothesis that channel m in primary network k is busy in time slot t. The conditional probability that channel m is available in primary network k, denoted by $a_m^k(t)$, can be derived as in (5.3), where θ_i^m represents a specific

$$\begin{aligned} a_m^k(t) &= \Pr(H_0 | W_i^m = \theta_i^m, \ i \in \mathcal{U}_m^k, \pi_m^k) \\ &= \frac{\Pr(W_i^m = \theta_i^m, i \in \mathcal{U}_m^k | H_0, \pi_m^k) \Pr(H_0 | \pi_m^k)}{\sum_{s \in \{0,1\}} \Pr(W_i^m = \theta_i^m, i \in \mathcal{U}_m^k | H_s, \pi_m^k) \Pr(H_s | \pi_m^k)} \\ &= \frac{\Pr(H_0 | \pi_m^k) \prod_{i \in \mathcal{U}_m^k} \Pr(W_i^m = \theta_i^m | H_0, \pi_m^k)}{\sum_{s \in \{0,1\}} \Pr(H_s | \pi_m^k) \prod_{i \in \mathcal{U}_m^k} \Pr(W_i^m = \theta_i^m | H_s, \pi_m^k)} \\ &= \frac{\Pr(H_0 | \pi_m^k) \prod_{i \in \mathcal{U}_m^k} \Pr(W_i^m = \theta_i^m | H_0)}{\sum_{s \in \{0,1\}} \Pr(H_s | \pi_m^k) \prod_{i \in \mathcal{U}_m^k} \Pr(W_i^m = \theta_i^m | H_s)} \\ &= \left[1 + \frac{\Pr(H_1 | \pi_m^k)}{\Pr(H_0 | \pi_m^k)} \prod_{i \in \mathcal{U}_m^k} \frac{\Pr(W_i^m = \theta_i^m | H_1)}{\Pr(W_i^m = \theta_i^m | H_0)} \right]^{-1} \\ &= \left[1 + \left(\varphi_m^k\right)^{u_m^k} \left(\phi_m^k\right)^{|\mathcal{U}_m^k| - u_m^k} \frac{\Pr(H_1 | \pi_m^k)}{\Pr(H_0 | \pi_m^k)} \right]^{-1}. \end{aligned} \tag{5.3}$$

sensing result (0 or 1), $\mathcal{U}_m^k$ is the subset of users in $\mathcal{U}^k$ (i.e., the set of CR nodes that are located within the coverage of primary network k) that sense channel m, u_m^k is the number of users in $\mathcal{U}_m^k$ observing channel m is idle, π_m^k represents the history of channel m in primary network k,[1] and φ_m^k and ϕ_m^k are defined as:

$$\begin{cases} \varphi_m^k = \frac{P(W_i^m=0|H_1)}{P(W_i^m=0|H_0)} = \frac{\delta_m}{1-\epsilon_m}, \text{ when } \theta_i^m = 0 \\ \phi_m^k = \frac{P(W_i^m=1|H_1)}{P(W_i^m=1|H_0)} = \frac{1-\delta_m}{\epsilon_m}, \text{ when } \theta_i^m = 1. \end{cases} \tag{5.4}$$

The third equality is due to independent sensing processes. The fourth equality is because sensing processes are independent of channel history. Based on the Markov chain channel model, we have (5.5), which can be recursively expanded:

$$\begin{cases} \Pr(H_0|\pi_m^k) = \lambda_m^k a_m^k(t-1) + \mu_m^k \left[1 - a_m^k(t-1)\right] \\ \Pr(H_1|\pi_m^k) = 1 - \Pr(H_0|\pi_m^k). \end{cases} \tag{5.5}$$

5.1.3 Spectrum Access

During the transmission phase of a time slot, a CR user determines which channel(s) to access for transmission of video data based on spectrum sensing results. Let κ_m^k be a threshold for spectrum access: channel m is considered idle if the estimate a_m^k is greater than the threshold, and busy otherwise. The availability of channel m in primary network k, denoted as A_m^k, is

$$A_m^k = \begin{cases} 0, & a_m^k \geq \kappa_m^k \\ 1, & \text{otherwise.} \end{cases} \tag{5.6}$$

For each channel m, we can calculate the probability of collision with primary users as:

$$\Pr(A_m^k = 0|H_1) = \sum_{i \in \psi_m^k} \binom{|\mathcal{U}_m^k|}{i} (1-\delta_m)^{|\mathcal{U}_m^k|-i} (\delta_m)^i, \tag{5.7}$$

where set ψ_m^k is defined as:

$$\psi_m^k = \left\{ i \,\middle|\, \left[1 + \varphi_m^i \phi_m^{|\mathcal{U}_m^k|-i} \frac{\Pr(H_1|\pi_m^k)}{\Pr(H_0|\pi_m^k)}\right]^{-1} \geq \kappa_m^k \right\}. \tag{5.8}$$

For nonintrusive spectrum access, the collision probability should be bounded with a prescribed threshold γ_m^k. A higher spectrum access threshold κ_m^k will reduce

[1] π_m^k represents the availability of channel m in primary network k in the previous time slot. If the channel was used in that time slot, π_m^k can be readily determined as 0 or 1, since the channel state was known (i.e., with or without ACKs). Otherwise, π_m^k can be estimated in the form of $a_m^k(t-1)$ as in (5.3).

the potential interference with primary users, but increase the chance of wasting transmission opportunities. For a given collision tolerance γ_m^k, we can solve $\Pr(A_m^k = 0|H_1) = \gamma_m^k$ for κ_m^k. The objective is to maximize CR users' spectrum access without exceeding the maximum collision probability with primary users.

Let $\Omega_{i,j}$ be the set of available channels at link $\{i, j\}$. Assuming $i \in \mathcal{U}^k$ and $j \in \mathcal{U}^{k'}$, we have

$$\Omega_{i,j} = \left\{ m \,\middle|\, A_m^k = 0 \text{ and } A_m^{k'} = 0 \right\}. \tag{5.9}$$

5.1.4 Link and Path Statistics

Due to the amplify-and-forward approach for video data transmission, there is no queueing delay at intermediate nodes. Assume each link has a fixed delay $\omega_{i,j}$ (i.e., processing and propagation delays). Let $\mathcal{P}_l^A$ be the set of all possible paths from z_l to d_l. For a given delay requirement T_{th}, the set of feasible paths $\mathcal{P}_l$ for video session l can be determined as:

$$\mathcal{P}_l = \left\{ \mathcal{P} \,\middle|\, \sum_{\{i,j\} \in \mathcal{P}} \omega_{i,j} \le T_{th},\ \mathcal{P} \in \mathcal{P}_l^A \right\}. \tag{5.10}$$

Let $p_{i,j}^m$ be the packet loss rate on channel m at link $\{i, j\}$. A packet is successfully delivered over link $\{i, j\}$ if there is no loss on all the channels that were used for transmitting the packet. The link loss probability $p_{i,j}$ can be derived as:

$$p_{i,j} = 1 - \prod_{m \in \mathcal{M}} (1 - p_{i,j}^m)^{I_m}, \tag{5.11}$$

where $\mathcal{M}$ is set of licensed channels and I_m is an indicator: $I_m = 1$ if channel m is used for the transmission, and $I_m = 0$ otherwise. Assuming independent link losses, the end-to-end loss probability for path $\mathcal{P}_l^h \in \mathcal{P}_l$ can be estimated as:

$$p_l^h = 1 - \prod_{\{i,j\} \in \mathcal{P}_l^h} (1 - p_{i,j}). \tag{5.12}$$

5.2 Problem Statement

We also aim to achieve fairness among the concurrent video sessions. It has been shown that *proportional fairness* can be achieved by maximizing the sum of logarithms of video PSNRs (i.e., utilities). Therefore, our objective is to maximize the overall system utility, i.e.,

$$\text{maximize:} \ \sum_l U_l(R_l) = \sum_l \log(Q_l(R_l)). \tag{5.13}$$

5.2.1 Multi-hop CR Network Video Streaming Problem

For the system described in Sect. 5.1, the problem of video over multi-hop CR networks consists of path selection for each video session and channel scheduling for each CR node along the chosen paths. We define two sets of index variables. For channel scheduling, we have

$$x_{i,j,m}^{l,h,r} = \begin{cases} 1, \ \text{at link } \{i, j\}, \text{ if channel } m \text{ is} \\ \quad \text{assigned to tunnel } r \text{ in path } \mathcal{P}_l^h \\ 0, \ \text{otherwise.} \end{cases} \tag{5.14}$$

For path selection, we have

$$y_l^h = \begin{cases} 1, \ \text{if video session } l \text{ selects path } \mathcal{P}_l^h \in \mathcal{P}_l \\ 0, \ \text{otherwise,} \end{cases} \tag{5.15}$$

Note that the indicators, $x_{i,j,m}^{l,h,r}$ and y_l^h, are not independent. If $y_l^h = 0$ for path $\mathcal{P}_l^h$, all the $x_{i,j,m}^{l,h,r}$'s on that path are 0. If link $\{i, j\}$ is not on path $\mathcal{P}_l^h$, all its $x_{i,j,m}^{l,h,r}$'s are also 0. For link $\{i, j\}$ on path $\mathcal{P}_l^h$, we can only choose those available channels in set $\Omega_{i,j}$ to schedule video transmission. That is, we have $x_{i,j,m}^{l,h,r} \in \{0, 1\}$ if $m \in \Omega_{i,j}$, and $x_{i,j,m}^{l,h,r} = 0$ otherwise. In the rest of the chapter, we use $\mathbf{x}$ and $\mathbf{y}$ to represent the vector forms of $x_{i,j,m}^{l,h,r}$ and y_l^h, respectively.

As discussed, the objective is to maximize the expected utility sum at the end of N_G time slots, as given in (5.13). Since $\log(Q_l(\mathrm{E}[R_l(0)]))$ is a constant, (5.13) is equivalent to the sum of utility increments of all the time slots, as

$$\begin{aligned} &\sum_l \log(Q_l(\mathrm{E}[R_l(N_G)])) - \log(Q_l(\mathrm{E}[R_l(0)])) \\ &= \sum_t \sum_l \{\log(Q_l(\mathrm{E}[R_l(t)])) - \log(Q_l(\mathrm{E}[R_l(t-1)]))\}. \end{aligned} \tag{5.16}$$

Therefore, (5.13) will be maximized if we maximize the expected utility increment during each time slot, which can be written as:

$$\sum_l \log(Q_l(\mathrm{E}[R_l(t)])) - \log(Q_l(\mathrm{E}[R_l(t-1)]))$$

$$= \sum_l \log\left(1 + \beta_l \frac{\mathrm{E}[R_l(t)] - \mathrm{E}[R_l(t-1)]}{Q_l(\mathrm{E}[R_l(t-1)])}\right)$$

$$= \sum_l \sum_{h \in \mathcal{P}_l} y_l^h \log\left(1 + \sum_r \sum_m \frac{\beta_l L_p x_{z_l, z_l', m}^{l,h,r}}{N_G T_s Q_l^{t-1}} (1 - p_{l,h}^r)\right)$$

$$= \sum_l \sum_{h \in \mathcal{P}_l} y_l^h \log\left(1 + \rho_l^t \sum_r \sum_m x_{z_l, z_l', m}^{l,h,r} (1 - p_{l,h}^r)\right),$$

where z_l' is the next hop from z_l on path $\mathcal{P}_l^h$, $p_{l,h}^r$ is the packet loss rate on tunnel r of path $\mathcal{P}_l^h$, $Q_l^{t-1} = Q_l(\mathrm{E}[R_l(t-1)])$, and $\rho_l^t = \beta_l L_p / (N_G T_s Q_l^{t-1})$.

From (5.11) and (5.12), the end-to-end packet loss rate for tunnel r on path $\mathcal{P}_l^h$ is:

$$p_{l,h}^r = 1 - \prod_{\{i,j\} \in \mathcal{P}_l^h} \prod_{m \in \mathcal{M}} (1 - p_{i,j}^m)^{x_{i,j,m}^{l,h,r}}. \tag{5.17}$$

We assume that each tunnel can only include one channel on each link. When there are multiple channels available at each link along the path, a CR source node can set up multiple tunnels to exploit the additional bandwidth. We then have the following constraint:

$$\sum_m x_{i,j,m}^{l,h,r} \le 1, \quad \forall \{i,j\} \in \mathcal{P}_l^h. \tag{5.18}$$

Considering availability of the channels, we further have,

$$\sum_r \sum_m x_{i,j,m}^{l,h,r} \le |\Omega_{i,j}|, \quad \forall \{i,j\} \in \mathcal{P}_l^h, \tag{5.19}$$

where $|\Omega_{i,j}|$ is the number of available channels on link $\{i,j\}$ defined in (5.9).

As discussed, each node is equipped with two transceivers: one for receiving and the other for transmitting video data during the transmission phase. Hence a channel cannot be used to receive and transmit data simultaneously at a relay node. We have for each channel m:

$$\sum_r x_{i,j,m}^{l,h,r} + \sum_r x_{j,k,m}^{l,h,r} \le 1, \quad \forall m, l, \forall h \in \mathcal{P}_l, \forall \{i,j\}, \{j,k\} \in \mathcal{P}_l^h. \tag{5.20}$$

Let n_l^h be the number of tunnels on path $\mathcal{P}_l^h$. For each source z_l and each destination d_l, the number of scheduled channels is equal to n_l^h. We have for each source node

$$\sum_r \sum_m x_{z_l, z_l', m}^{l,h,r} = n_l^h y_l^h, \quad \forall h \in \mathcal{P}_l, \forall l. \tag{5.21}$$

Let d'_l be the last hop to destination d_l on path $\mathcal{P}_l^h$, we have for each destination node

$$\sum_r \sum_m x_{d'_l, d_l, m}^{l,h,r} = n_l^h y_l^h, \quad \forall\, h \in \mathcal{P}_l, \forall\, l. \tag{5.22}$$

At a relay node, the number of channels used to receive data is equal to that of channels used to transmit data, due to flow conservation and amplify-and-forward. At relay node j for session l, assume $\{i, j\} \in \mathcal{P}_l^h$ and $\{j, k\} \in \mathcal{P}_l^h$. We have,

$$\sum_r \sum_m x_{i,j,m}^{l,h,r} = \sum_r \sum_m x_{j,k,m}^{l,h,r}, \quad \forall\, h \in \mathcal{P}_l, \forall\, l, \forall\, \{i, j\}, \{j, k\} \in \mathcal{P}_l^h. \tag{5.23}$$

We also consider hardware-related constraints on path selection. We summarize such constraints in the following general form for ease of presentation:

$$\sum_l \sum_{h \in \mathcal{P}_l} w_{l,h}^g y_l^h \le 1, \forall\, g. \tag{5.24}$$

To simplify exposition, we choose at most one path in $\mathcal{P}_l$ for video session l. Such a single path routing constraint can be expressed as $\sum_h y_l^h \le 1$, which is a special case of (5.24) where $w_{l,h}^1 = 1$ for all h, and $w_{l',h}^g = 0$ for all $g \ne 1, l' \ne l$, and h. We can also have $\sum_h y_l^h \le \xi$ to allow up to ξ paths for each video session. In order to achieve optimality in the general case of multi-path routing, an optimal scheduling algorithm should be designed to dispatch packets to paths with different conditions (e.g., different number of tunnels and delays).

There are also disjointedness constraints for the chosen paths. This is because each CR node is equipped with two transceivers and both will be used for a video session if it is included in a chosen path. Such disjointedness constraint is also a special case of (5.24) with the following definition for $w_{l,h}^g$ for each CR node g:

$$w_{l,h}^g = \begin{cases} 1, \text{ if node } g \in \text{ path } \mathcal{P}_l^h \\ 0, \text{ otherwise,} \end{cases} \tag{5.25}$$

Finally, we formulate the problem of multi-hop CR network video streaming (OPT-CRV) as:

$$\text{maximize: } \sum_l \sum_{h \in \mathcal{P}_l} y_l^h \log\left(1 + \rho_l^t \sum_r \sum_m x_{z_l, z'_l, m}^{l,h,r} (1 - p_{l,h}^r)\right) \tag{5.26}$$

subject to: (5.14)–(5.24).

Algorithm 1: Sequential Fixing Algorithm (SF) for Problem OPT-CRV ([4], © 2010 IEEE)

1 Relax integer variables $x_{i,j,m}^{l,h,r}$, y_l^h, and n_l^h ;
2 Solve the relaxed problem using a constrained NLP solver ;
3 **if** (*there is* y_l^h *not fixed*) **then**
4   Find the largest $y_{l'}^{h'}$, where $[l', h'] = \arg\max\{y_l^h\}$, and fix it to 1 ;
5   Fix other y_l^h's according to constraint (5.24) ;
6   Go to Step 2 ;
7 **end**
8 **if** (*there is* $x_{i,j,m}^{l,h,r}$ *not fixed*) **then**
9   Find the largest $x_{i',j',m'}^{l',h',r'}$, where $[i', j', m', l', h', r'] = \arg\max\{x_{i,j,m}^{l,h,r}\}$, and set it to 1 ;
10   Fix other $x_{i,j,m}^{l,h,r}$'s according to the constraints ;
11   **if** (*there is other variable that is not fixed*) **then**
12   Go to Step 2 ;
13   **else**
14   Fix n_l^h's based on $\mathbf{x}$ and $\mathbf{y}$;
15   Exit with feasible solution $\{\mathbf{x}, \mathbf{y}, \mathbf{n}\}$;
16   **end**
17 **end**

5.2.2 Centralized Algorithm and Upper/Lower Bounds

Problem OPT-CRV is in the form of MINLP (without continuous variables), which is NP-hard in general. We first describe a centralized algorithm to derive performance bounds in this section, and then present a distributed algorithm based on dual decomposition in the next section.

We first obtain a relaxed *nonlinear programming* (NLP) version of OPT-CRV. The binary variables $x_{i,j,m}^{l,h,r}$ and y_l^h are relaxed to take values in [0,1]. The integer variables n_l^h are treated as nonnegative real numbers. It can be shown that the relaxed problem has a concave object function and the constraints are convex. This relaxed problem can be solved using a constrained nonlinear optimization problem solver. If all the variables are integer in the solution, then we have the exact optimal solution. Otherwise, we obtain an infeasible solution, which produces an upper bound for the problem. This is given in Lines 17 $\sim$ 2 in Algorithm 1.

We also develop a *sequential fixing algorithm* (SF) for solving OPT-CRV. The pseudo-code is given in Algorithm 1. SF iteratively solves the relaxed problem, fixing one or more integer variables after each iteration [1,17]. In Algorithm 1, Lines 3 $\sim$ 7 fix the path selection variables y_l^h, and Lines 8 $\sim$ 17 fix the channel scheduling variables $x_{i,j,m}^{l,h,r}$ and tunnel variables n_l^h. The tunnel variables n_l^h can be computed using (5.21) after $x_{i,j,m}^{l,h,r}$ and y_l^h are solved. When the algorithm terminates, it produces a feasible solution that yields a lower bound for the objective value.

5.3 Dual Decomposition

SF is a centralized algorithm requiring global information. It may not be suitable for multi-hop wireless networks, although the upper and lower bounds provide useful insights into the performance limits. In this section, we develop a distributed algorithm for Problem OPT-CRV and analyze its optimality and convergence performance.

5.3.1 Decompose Problem OPT-CRV

Since the domains of $x^{l,h,r}_{i,j,m}$ defined in (5.18)–(5.23) for different paths do not intersect with each other, we can decompose Problem OPT-CRV into two subproblems. The first subproblem deals with channel scheduling for maximizing the expected utility on a chosen path $\mathcal{P}_l^h$. We have the *channel scheduling* problem (OPT-CS) as:

$$H_l^h = \max_{\mathbf{x}} \sum_r \sum_m x^{l,h,r}_{z_l,z'_l,m}(1 - p^r_{l,h}) \tag{5.27}$$

subject to:

$$(5.18)\text{–}(5.23),\ x^{l,h,r}_{z_l,z'_l,m} \in \{0, 1\},\ \text{for all } l, h, r, m.$$

In the second part, optimal paths are selected to maximize the overall objective function. Letting $F_l^h = \log\left(1 + \rho_l^T H_l^h\right)$, we have the following *path selection* problem (OPT-PS):

$$\text{maximize:}\ f(\mathbf{y}) = \sum_l \sum_h F_l^h y_l^h \tag{5.28}$$

subject to:

$$\sum_l \sum_{h \in \mathcal{P}_l} w^g_{l,h} y_l^h \le 1,\ \text{for all } g$$

$$y_l^h \in \{0, 1\},\ \text{for all } l, h.$$

5.3.2 Solve the Channel Scheduling Subproblem

We have the following result for assigning available channels at a relay node.

Theorem 5.1 *Consider three consecutive nodes along a path, denoted as nodes i, j, and k. Idle channels* 1 *and* 2 *are available at link $\{i, j\}$ and idle channels* 3 *and* 4 *are available at link $\{j, k\}$. Assume the packet loss rates of the four channels satisfy*

$p_{i,j}^1 > p_{i,j}^2$ and $p_{j,k}^3 > p_{j,k}^4$. *To set up two tunnels, assigning channels* $\{1, 3\}$ *to one tunnel and channels* $\{2, 4\}$ *to the other tunnel achieves the maximum expectation of successful transmission on path section* $\{i, j, k\}$.

Proof Let the success probabilities on the channels be $\tilde{p}_{i,j}^1 = 1 - p_{i,j}^1$, $\tilde{p}_{i,j}^2 = 1 - p_{i,j}^2$, $\tilde{p}_{j,k}^3 = 1 - p_{j,k}^3$, and $\tilde{p}_{j,k}^4 = 1 - p_{j,k}^4$. We have $\tilde{p}_{i,j}^1 < \tilde{p}_{i,j}^2$ and $\tilde{p}_{j,k}^3 < \tilde{p}_{j,k}^4$. Comparing the success probabilities of the channel assignment given in Theorem 5.1 and that of the alternative assignment, we have $\tilde{p}_{i,j}^1 \tilde{p}_{j,k}^3 + \tilde{p}_{i,j}^2 \tilde{p}_{j,k}^4 - \tilde{p}_{i,j}^1 \tilde{p}_{j,k}^4 - \tilde{p}_{i,j}^2 \tilde{p}_{j,k}^3 = (\tilde{p}_{i,j}^1 - \tilde{p}_{i,j}^2)(\tilde{p}_{j,k}^3 - \tilde{p}_{j,k}^4) > 0$. The result follows. □

According to Theorem 5.1, a greedy approach, which always chooses the channel with the lowest loss rate at each link when setting up tunnels along a path, produces the optimal overall success probability. More specifically, when there is only one tunnel to be set up along a path, the tunnel should consist of the most reliable channels available at each link along the path. When there are multiple tunnels to set up along a path, tunnel 1 should consist of the most reliable channels that are available at each link; tunnel 2 should consist of the second most reliable links available at each link; and so forth.

Define the set of loss rates of the available channels on link $\{i, j\}$ as $\Lambda_{i,j} = \{p_{i,j}^m | m \in \Omega_{i,j}\}$. The greedy algorithm is given in Algorithm 2, with which each video source node solves Problem OPT-CS for each feasible path. Lines 2 in Algorithm 2 checks if there are more channels to assign and the algorithm terminates if no channel is left. In Lines 4 $\sim$ 10, links with only one available channel are assigned to tunnel r and the neighboring links with the same available channels are removed due to constraint (5.20). In Lines 12 $\sim$ 14, links with more than two channels are grouped to be assigned later. In Lines 16 $\sim$ 17, the available channel with the lowest packet loss rate is assigned to tunnel r at each unallocated link, according to Theorem 5.1. To avoid co-channel interference, the same channel on neighboring links is removed as in Lines 18 $\sim$ 26.

5.3.3 Solve the Path Selection Subproblem

To solve Problem OPT-PS, we first relax binary variables y_l^h to allow them take real values in [0,1] and obtain the following *relaxed path selection* problem (OPT-rPS):

$$\text{maximize:} \quad f(\mathbf{y}) = \textstyle\sum_l \sum_h F_l^h y_l^h \tag{5.29}$$

$$\text{subject to:}$$

$$\textstyle\sum_l \sum_{h \in \mathcal{P}_l} w_{l,h}^g y_l^h \le 1, \ \text{for all } g$$

$$0 \le y_l^h \le 1, \ \text{for all } h, l.$$

We then introduce positive Lagrange Multipliers e_g for the path selection constraints in Problem OPT-rPS and obtain the corresponding *Lagrangian function*:

Algorithm 2: Greedy Algorithm for Channel Scheduling ([4], © 2010 IEEE)

1 Initialization: tunnel $r = 1$, link $\{i, j\}$'s from z_l to d_l ;
2 **if** $(|\Lambda_{i,j}| == 0)$ **then** Exit ;
3 **if** $(|\Lambda_{i,j}| == 1)$ **then**
4   Assign the single channel in $\Lambda_{i,j}$, m', to tunnel r ;
5   Check neighboring link $\{k, i\}$;
6   **if** $\left(p_{k,i}^{m'} \in \Lambda_{k,i}\right)$ **then**
7   Remove $p_{k,i}^{m'}$ from $\Lambda_{k,i}$, $i \leftarrow k$, $j \leftarrow i$, and go to Step 2 ;
8   **else**
9   Go to Step 12 ;
10   **end**
11 **else**
12   Put $\Lambda_{i,j}$ in set Λ_l^h ;
13   **if** (*node j is not destination* d_l) **then** $i \leftarrow j$, $j \leftarrow v$, and go to Step 2 ;
14 **end**
15 **while** $(\Lambda_l^h \neq \emptyset)$ **do**
16   Find the maximum value $p_{i',j'}^{m'}$ in set Λ_l^h, $\{i', j', m'\} = \arg\min\{p_{i,j}^m\}$;
17   Assign channel m' to tunnel r, and remove set $\Lambda_{i',j'}$ from set Λ_l^h ;
18   Check neighboring link $\{k, i\}$ and $\{j, v\}$;
19   **if** $\left(p_{k,i}^{m'} \in \Lambda_{k,i} and\ \Lambda_{k,i} \in \Lambda_l^h\right)$ **then**
20   Remove $p_{k,i}^{m'}$ from $\Lambda_{k,i}$;
21   **if** $(\Lambda_{k,i} = \emptyset)$ **then** Exit;
22   **end**
23   **if** $\left(p_{j,v}^{m'} \in \Lambda_{j,v} and\ \Lambda_{j,v} \in \Lambda_l^h\right)$ **then**
24   Remove $p_{j,v}^{m'}$ from $\Lambda_{j,v}$;
25   **if** $(\Lambda_{j,v} = \emptyset)$ **then** Exit ;
26   **end**
27 **end**
28 Compute the next tunnel: $r \leftarrow r + 1$ and go to Step 2 ;

$$
\begin{aligned}
\mathcal{L}(\mathbf{y}, \mathbf{e}) &= \sum_l \sum_h F_l^h y_l^h + \sum_g e_g (1 - \sum_l \sum_h w_{l,h}^g y_l^h) \\
&= \sum_l \sum_h (F_l^h y_l^h - \sum_g w_{l,h}^g y_l^h e_g) + \sum_g e_g \\
&= \sum_l \sum_h \mathcal{L}_l^h (y_l^h, \mathbf{e}) + \sum_g e_g .
\end{aligned} \tag{5.30}
$$

Problem (5.30) can be decoupled since the domains of y_l^h's do not overlap. Relaxing the coupling constraints, it can be decomposed into two levels. At the lower level, we have the following subproblems, one for each path $\mathcal{P}_l^h$,

$$
\max_{0 \leq y_l^h \leq 1} \mathcal{L}_l^h (y_l^h, \mathbf{e}) = F_l^h y_l^h - \sum_g w_{l,h}^g y_l^h e_g . \tag{5.31}
$$

Algorithm 3: Distribution Algorithm for Path Selection ([4], © 2010 IEEE)

1 Initialization: set $\tau = 0$, $e_g(0) > 0$ and step size $s \in [0, 1]$;
2 Each source locally solves the lower level problem in (5.31) ;
3 **if** $\left(F_l^h - \sum_g d_{l,h}^g e_g(\tau)) > 0\right)$ **then**
4 | $y_l^h = y_l^h + s$, $y_l^h = \min\{y_l^h, 1\}$;
5 **else**
6 | $y_l^h = y_l^h - s$, $y_l^h = \max\{y_l^h, 0\}$;
7 **end**
8 Broadcast solution $y_l^h(\mathbf{e}(\tau))$;
9 Each source updates $\mathbf{e}$ according to (5.33) and broadcasts $\mathbf{e}(\tau + 1)$ through the common control channel ;
10 $\tau \leftarrow \tau$+1 and go to Step 2 until termination criterion is satisfied ;

At the higher level, by updating the dual variables e_g, we can solve the *relaxed dual problem*:

$$\min_{\mathbf{e}\geq 0} \; q(\mathbf{e}) = \sum_l \sum_h \mathcal{L}_l^h \left(\left(y_l^h\right)^*, \mathbf{e} \right) + \sum_g e_g, \tag{5.32}$$

where $\left(y_l^h\right)^*$ is the optimal solution to (5.31). Since the solution to (5.31) is unique, the relaxed dual problem (5.32) can be solved using the following *subgradient method* that iteratively updates the Lagrange Multipliers [53]:

$$e_g(\tau + 1) = \left[e_g(\tau) - \alpha(\tau)(1 - \sum_l \sum_h w_{l,h}^g y_l^h) \right]^+, \tag{5.33}$$

where τ is the iteration index, $\alpha(\tau)$ is a sufficiently small positive step size and $[x]^+$ denotes $\max\{x, 0\}$. The pseudo code for the distributed algorithm is given in Algorithm 3.

5.3.4 Optimality and Convergence Analysis

The distributed algorithm in Algorithm 3 iteratively updates the dual variables until they converge to stable values. In this section, we first prove that the solution obtained by the distributed algorithm is also optimal for the original path selection problem OPT-PS. We then derive the convergence condition for the distributed algorithm.

Fact 1 ([53]). *Consider a linear problem involving both equality and inequality constraints*

$$\begin{aligned} \text{maximize: } & \mathbf{a}'\mathbf{x} \\ \text{subject to: } & \mathbf{h}_1'\mathbf{x} = b_1, \ \ldots, \ \mathbf{h}_m'\mathbf{x} = b_m \\ & \mathbf{g}_1'\mathbf{x} \le c_1, \ \ldots, \ \mathbf{g}_r'\mathbf{x} \le c_r, \end{aligned} \tag{5.34}$$

where $\mathbf{a}$, $\mathbf{h}_i$, and $\mathbf{g}_j$ *are column vectors in* $\mathcal{R}_n$, b_i's *and* c_j's *are scalars, and* $\mathbf{a}'$ *is the transpose of* $\mathbf{a}$. *For any feasible point* $\mathbf{x}$, *the set of active inequality constraints is denoted by* $\mathcal{A}(\mathbf{x}) = \left\{ j | \mathbf{g}_j'\mathbf{x} = c_j \right\}$. *If* $\mathbf{x}^*$ *is a maximizer of inequality constrained problem* (5.34), $\mathbf{x}^*$ *is also a maximizer of the following equality constrained problem*:

$$\begin{aligned} \text{maximize: } & \mathbf{a}'\mathbf{x} \\ \text{subject to: } & \mathbf{h}_1'\mathbf{x} = b_1, \ \ldots, \ \mathbf{h}_m'\mathbf{x} = b_m \\ & \mathbf{g}_j'\mathbf{x} = c_j, \forall \ j \in \mathcal{A}(\mathbf{x}). \end{aligned} \tag{5.35}$$

Lemma 5.1 *The optimal solution for the relaxed primal problem OPT-rPS in* (5.29) *is also feasible and optimal for the original Problem OPT-PS in* (5.28).

Proof According to Fact 1, the linearized problem of OPT-PS, i.e., OPT-rPS, can be rewritten as an equality constrained problem in the following form:

$$\text{maximize: } \mathbf{F}'\mathbf{y} \tag{5.36}$$

$$\begin{aligned} \text{subject to: } & \mathbf{w}_j'\mathbf{y} = 1, \quad j \in \mathcal{A}(\mathbf{y}^*) \\ & 0 \le y_l^h \le 1, \ \text{ for all } h, l, \end{aligned} \tag{5.37}$$

where $\mathbf{F}$, $\mathbf{w}_j$'s, and $\mathbf{y}$ are column vectors with elements F_l^h, $w_{l,h}^g$, and y_l^h, respectively. We apply *Gauss-Jordan elimination* to the constraints in (5.37) to solve for $\mathbf{y}$. Since there is not sufficient number of equations, some y_l^h's are free variables (denoted as y_i^f) and the rest are dependent variables (denoted as y_j^d). Assuming there are r free variables, the dependent variables can be written as linear combinations of the free variables after Gauss-Jordan elimination, as

$$y_j^d = \sum_{i=1}^{r} \bar{w}_j^i y_i^f + \bar{b}_j, \ \ j \in \mathcal{A}(y_i^*). \tag{5.38}$$

Due to Gauss-Jordan elimination and binary vectors $\mathbf{w}_j$'s, $\bar{w}_j^i$ and $\bar{b}_j$ in (5.38) are all integers. Therefore, if all the free variables y_i^f attain binary values, then all the dependent variables y_j^d computed using (5.38) will also be integers. Since $0 \le y_j^d \le 1$, being integers means that they are either 0 or 1, i.e., binaries. That is, such a solution will be feasible.

Next we substitute (5.38) into problem (5.36) to eliminate all the dependent variables. Then we obtain a unconstrained problem with only r free variables, as

$$\text{maximize:} \quad \sum_{i=1}^{r} \bar{F}_i y_i^f + \bar{b}_0 \tag{5.39}$$

Since the free variables y_i^f's take value in $\{0, 1\}$, this problem can be easily solved as follows. If the coefficient $\bar{F}_i > 0$, we set $y_i^f = 1$; otherwise, if $\bar{F}_i < 0$, we set $y_i^f = 0$. Thus (5.39) achieves its maximum objective value. Once all the free variables are determined with their optimal binary values, we compute the dependent variables using (5.38), which are also binary as discussed above. Thus we obtain a feasible solution, which is optimal.

Lemma 5.2 *If the relaxed primal Problem OPT-rPS in* (5.29) *has an optimal solution, then the relaxed dual problem* (5.32) *also has an optimal solution and the corresponding optimal values of the two problems are identical.*

Proof By definition, the problems in (5.30) and (5.32) are primal/dual problems. The primal problem always has an optimal solution because it is bounded. Since Problem OPT-rPS is an LP problem, the relaxed dual problem is also bounded and feasible. Therefore the relaxed dual problem also has an optimal solution. We have the *strong duality* if the primal problem is convex, which is the case here since Problem OPT-rPS is an LP problem. □

We have Theorem 5.2 on the optimality of the path selection solution, which follows naturally from Lemmas 5.1 and 5.2.

Theorem 5.2 *The optimal solution to the relaxed dual problem* (5.31) *and* (5.32) *is also feasible and optimal to the original path selection Problem* OPT-PS *given in* (5.28).

As discussed, the relaxed dual problem (5.32) can be solved using the *subgradient method* that iteratively updates the Lagrange Multipliers. We have the following theorem on the convergence of the distributed algorithm given in Algorithm 3.

Theorem 5.3 *Let* $\mathbf{e}^*$ *be the optimal solution. The distributed algorithm in Algorithm* 13 *converges if the step sizes* $\alpha(\tau)$ *in* (5.33) *satisfy the following condition*:

$$0 < \alpha(\tau) < \frac{2\left[q(\mathbf{e}(\tau)) - q(\mathbf{e}^*)\right]}{||G(\tau)||^2}, \quad \text{for all } \tau, \tag{5.40}$$

where $G(\tau)$ *is the gradient of* $q(\mathbf{e}(\tau))$.

Proof Since $q(\mathbf{e}(\tau))$ is a linear function, we have subgradient equality, as

$$q(\mathbf{e}(\tau)) - q(\mathbf{e}^*) = \left[\mathbf{e}(\tau) - \mathbf{e}^*\right]' G(\tau).$$

It then follows that

$$\begin{aligned}
&||\mathbf{e}(\tau) - \alpha(\tau)G(\tau) - \mathbf{e}^*||^2 \\
&= ||\mathbf{e}(\tau)-\mathbf{e}^*||^2 - 2\alpha(\tau)[\mathbf{e}(\tau)-\mathbf{e}^*]'G(\tau) + (\alpha(\tau))^2||G(\tau)||^2 \\
&= ||\mathbf{e}(\tau)-\mathbf{e}^*||^2 - 2\alpha(\tau)[q(\mathbf{e}(\tau))-q(\mathbf{e}^*)] + (\alpha(\tau))^2||G(\tau)||^2.
\end{aligned} \tag{5.41}$$

If $\alpha(\tau)$ satisfy (5.40), the sum of the last two terms in (5.41) is negative. It follows that, $||\mathbf{e}(\tau) - \alpha(\tau)G(\tau) - \mathbf{e}^*|| < ||\mathbf{e}(\tau) - \mathbf{e}^*||$. Since the projection operation is *nonexpansive*, we have,

$$\begin{aligned}
&||\mathbf{e}(\tau+1) - \mathbf{e}^*|| \\
&= ||[\mathbf{e}(\tau) - \alpha(\tau)G(\tau)]^+ - [\mathbf{e}^*]^+|| \\
&\leq ||\mathbf{e}(\tau) - \alpha(\tau)G(\tau) - \mathbf{e}^*|| \\
&< ||\mathbf{e}(\tau) - \mathbf{e}^*||,
\end{aligned}$$

which states the conditional convergence of the algorithm. □

Since the optimal solution $\mathbf{e}^*$ is not known a priori, we use the following approximation in the algorithm:

$$\alpha(\tau) = \frac{q(\mathbf{e}(\tau)) - \hat{q}(\tau)}{||G(\tau)||^2}, \tag{5.42}$$

where $\hat{q}(\tau)$ is the current estimate for $q(\mathbf{e}^*)$. We choose the mean of the objective values of the relaxed primal and dual problems for $\hat{q}(\tau)$.

5.3.5 Practical Considerations

Our distributed algorithms are based on the fact that the computation is distributed on each feasible path. The OPT-CS algorithm requires information on channel availability and packet loss rates at the links of feasible paths. The OPT-PS algorithm computes the primal variable y_l^h for each path and broadcasts Lagrangian multipliers over the control channel to all the source nodes. We assume a perfect control channel such that channel information can be effectively distributed and shared, which is not confined by the time slot structure [31].

We assume relatively large timescales for the primary network time slots, and small to medium diameter for the CR network, such that there is sufficient time for timely feedback of channel information to the video source nodes and for the distributed algorithms to converge. Otherwise, channel information can be estimated using (5.5) based on delayed feedback, leading to suboptimal solutions. If the time slot is too short, the distributed algorithm may not converge to the optimal solution (see Fig. 5.7). We focus on developing the CR video framework in this chapter, and will investigate these issues in our future work.

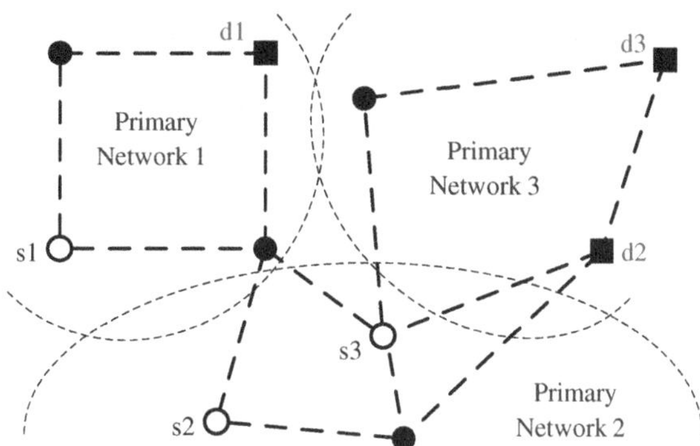

Fig. 5.3 The topology of the multi-hop CR network used in simulations. Note that only the video source nodes, video destination nodes, and those nodes along the precomputed paths are shown in the topology ([4], ©2010 IEEE)

5.4 Simulation Results

5.4.1 Methodology and Simulation Settings

We implement the proposed algorithms with a combination of C and MATLAB (i.e., for solving the relaxed NLP problems), and evaluate their performance with simulations. For the results reported in this section, we have $K = 3$ primary networks and $M = 10$ channels. There are 56, 55, and 62 CR users in the coverage areas of primary networks 1, 2, and 3, respectively. The $|\mathcal{U}_m^1|$'s are [5 4 6 4 8 7 5 6 7 4] (i.e., five users sense channel 1, four users sense channel 2, and so forth); the $|\mathcal{U}_m^2|$'s are [4 6 5 7 6 5 3 8 5 6], and the $|\mathcal{U}_m^3|$'s are [8 6 5 4 7 6 8 5 6 7]. The topology is shown in Fig. 5.3.

We choose $L_p = 100$, $T_s = 0.02$ and $N_G = 10$. The channel utilization is $\eta_m^k = 0.6$ for all the channels. The probability of false alarm is $\epsilon_m^k = 0.3$ and the probability of miss detection is $\delta_m^k = 0.2$ for all m and k, unless otherwise specified. Channel parameters λ_m^k and μ_m^k are set between (0, 1). The maximum allowed collision probability γ_m^k is set to 0.2 for all the M channels in the three primary networks.

We consider three video sessions, each streaming a video in the Common Intermediate Format (CIF, 352 × 288), i.e., *Bus* to destination 1, *Foreman* to destination 2, and *Mother & Daughter* to destination 3. The frame rate is 30 fps, and a GOP consists of 10 frames. We assume that the duration of a time slot is 0.02 s and each GOP should be delivered in 0.2 s (i.e., 10 time slots).

We compare four schemes in the simulations:

- The upper bounding solution by solving the relaxed version of Problem OPT-CRV using an NLP solver.
- The proposed distributed algorithm in Algorithms 2 and 3.
- The sequential fixing algorithm given in Algorithm 1, which computes a lower bounding solution.

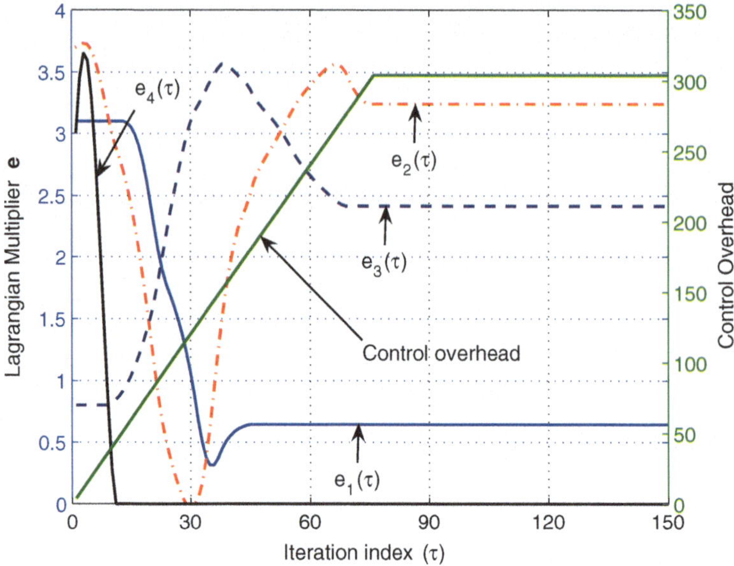

Fig. 5.4 Illustrate the convergence of the distributed algorithm ([4], ©2010 IEEE)

- A greedy heuristic where at each hop, the link with the most available channels is used.

Each point in the figures is the average of 10 simulation runs, with 95 % confidence intervals plotted as error bars in the figures. The 95 % confidence intervals are negligible in all the figures.

5.4.2 Simulation Results

Algorithm Performance

To demonstrate the convergence of the distributed algorithm, we plot the traces of the four Lagrangian multipliers in Fig. 5.4. We observe that all the Lagrangian multipliers converge to their optimal values after 76 iterations. We also plot the control overhead as measured by the number of distinct broadcast messages for $e_i(\tau)$ using the y-axis on the right-hand side. The overhead curve increases linearly with the number of iterations and gets flat (i.e., no more broadcast message) when all the Lagrangian multipliers converge to their optimal values.

We examine the impact of spectrum sensing errors in Fig. 5.5. We test six sensing error combinations $\{\epsilon_m, \delta_m\}$ as follows: $\{0.1, 0.5\}$, $\{0.2, 0.3\}$, $\{0.3, 0.2\}$, $\{0.5, 0.11\}$, $\{0.7, 0.06\}$, and $\{0.9, 0.02\}$, and plot the average PSNR values of the Foreman session. It is interesting to see that the best video quality is achieved when the false

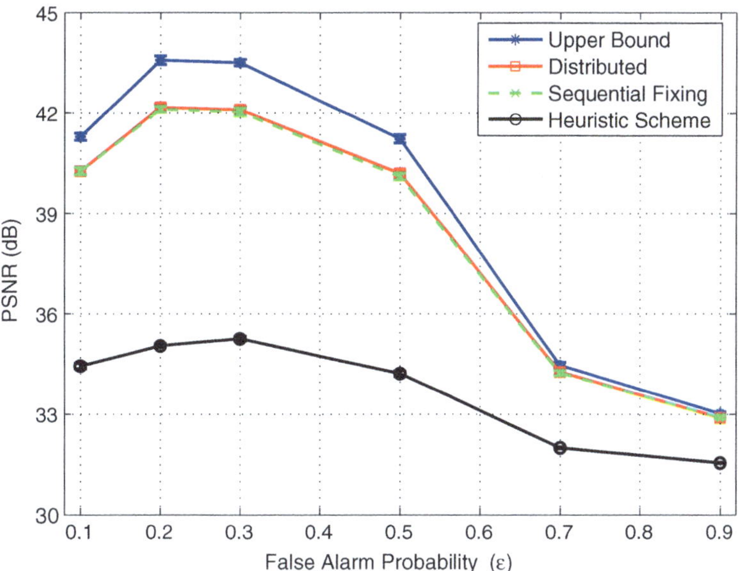

Fig. 5.5 Video PSNRs versus spectrum sensing error ([4], © 2010IEEE)

alarm probability ϵ_m is between 0.2 and 0.3. Since the two error probabilities are correlated, increasing one will generally decrease the other. With a larger ϵ_m, CR users are more likely to waste spectrum opportunities that are actually available, leading to lower bandwidth for videos and poorer video quality, as shown in Fig. 5.5. On the other hand, a larger δ_m implies more aggressive spectrum access and more severe interference to primary users. Therefore when ϵ_m is lower than 0.2 (and δ_m is higher than 0.3), the CR nodes themselves also suffer from the collisions and the video quality degrades.

Impact of Primary Network Parameters

In Fig. 5.6, we examine the impact of channel utilization η on received video quality. We focus on Session 2 with the Foreman sequence. The average PSNRs achieved by the four schemes are plotted when η is increased from 0.6 to 0.9 for all licensed channels. Intuitively, a smaller η allows more transmission opportunities for CR nodes, leading to improved video quality. This is illustrated in the figure where all the four curves decrease as η gets larger. The distributed scheme achieves PSNRs very close to that obtained by sequential fixing, and both of them are close to the upper bound. The heuristic scheme is inefficient in exploiting the available spectrum even when the channel utilization is low.

As discussed, the time slot duration is also an important parameter that may affect the convergence of the distributed algorithm. In Fig. 5.7, we keep the same

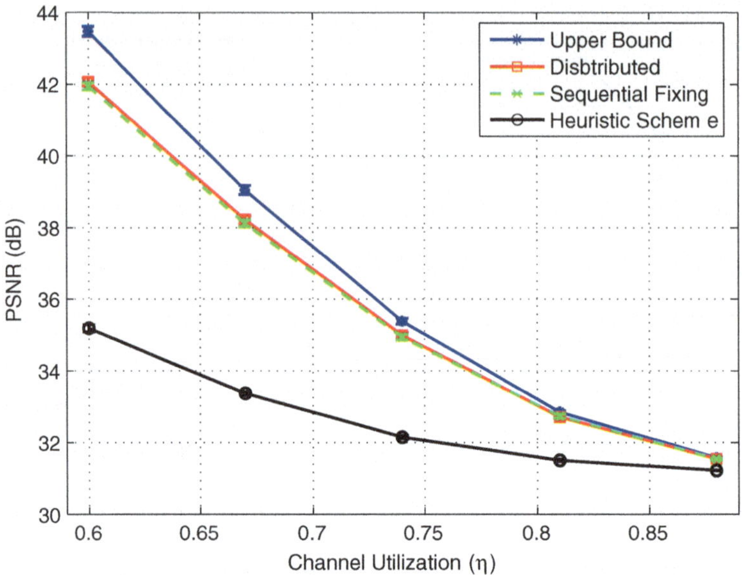

Fig. 5.6 Video PSNRs versus primary user channel utilization η ([4], ©2010 IEEE)

network and video session settings, while increasing the time slot duration as 4 ms, 10 ms, 20 ms, 40 ms and 100 ms. For a given time slot duration, we let the distributed algorithm run for 5 % of the time slot duration, starting from the beginning of the time slot, and then stop. The solution that the algorithm produces when it is stopped will be used for video transmission in the remainder of this time slot. It can be seen that when the time slot is 4 ms, the algorithm does not converge after 5 %×4=0.2 ms, and the PSNR produced by the distributed algorithm is low (but still higher than that of the heuristic algorithm). When the time slot duration is sufficiently large (e.g., over 10 ms), the algorithm can converge and the proposed algorithm produces very good video quality as compared to the upper bound and the lower bound given by the sequential fixing algorithm.

Comparison of MPEG-4 FGS and H.264/SVC MGS Videos

Finally, we compare MPEG-4 FGS and H.264/SVC MGS videos, while keeping the same settings. It has been shown that H.264/SVC has better rate-distortion performance than MPEG-4 FGS due to the use of efficient hierarchical prediction structures, the inter-layer prediction mechanisms, improved drift control mechanism, and the efficient coding scheme in H.264/AVC [57]. Although MGS has Network Abstraction Layer (NAL) unit-based granularity, it achieves similar rate-distortion performance as H.264/SVC FGS [57].

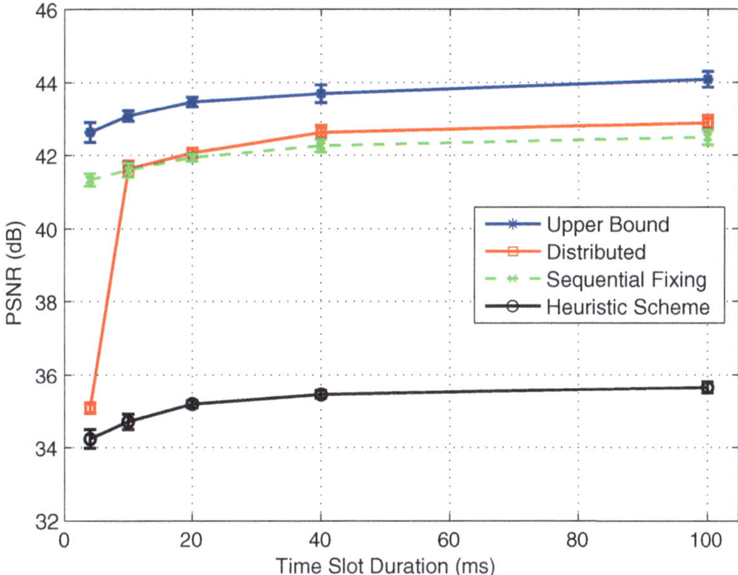

Fig. 5.7 Impact of time slot duration on received video quality ([4], ©2010 IEEE)

We plot the upper bounds and the distributed algorithm results in Figs. 5.8 and 5.9 for various channel utilizations and false alarm probabilities, respectively. From the figures, it can be observed that there is a gap of about 2.5 dB between the H.264/SVC MGS and MPEG-4 FGS curves, which clearly demonstrates the rate-distortion efficiency of MGS over MPEG-4 FGS. The proposed algorithm can effectively handle both MGS and FGS videos, and the same trend is observed in both cases.

5.5 Related Work

The high potential of CRs has attracted considerable interest from the wireless community [12, 13, 38]. The mainstream CR research has been focused on spectrum sensing and dynamic spectrum access issues [18, 31, 84–86]. Several papers have addressed the impact of spectrum sensing errors on the design of spectrum access schemes [23, 86–88]. The approach of iteratively sensing a selected subset of available channels has been adopted in the design of CR MAC protocols (e.g., see [18, 31 ,86]). Our work is complementary to this class of work by providing an important application for the enhanced spectrum efficiency achieved by spectrum sensing and access schemes reported in the literature.

Multi-hop SDR or CR networks have been studied in a few recent works [17, 89, 90]. The authors formulate cross-layer optimization problems considering factors from the PHY up to the transport layer. Distributed algorithms are developed by applying the dual decomposition technique [53, 91]. We adopt similar methodology

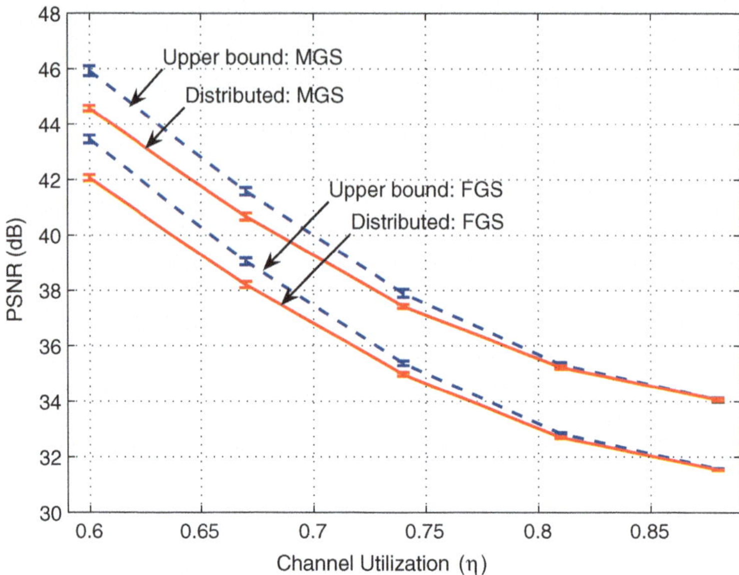

Fig. 5.8 Comparison of MPEG-4 FGS video with H.264/SVC MGS video under various channel utilizations ([4], ©2010 IEEE)

in this chapter, but address the more challenging problem of streaming real-time videos.

The important problem of QoS provisioning in CR networks has been considered in a few papers [31, 34, 87], where the focus is still on the so-called network-centric metrics such as maximum throughput and delay [31, 87]. In a recent work [87], Urgaonkar and Neely derive an interesting delay throughput trade-off for a multi-cell cognitive radio network, while primary user protection is achieved by stabilizing a virtual "collision queue."

In [34], a game-theoretic framework is described for resource allocation for multimedia transmissions in spectrum agile wireless networks. In this interesting work, each wireless station plays a resource management game, which is coordinated by a network moderator. A mechanism-based resource management scheme determines the amount of transmission time to be allocated to various users on different frequency bands such that certain global system metrics are optimized. In our prior work [1], we consider video multicast in an infrastructure-based CR network. We present effective greedy heuristic algorithms for scheduling video data, with proved optimality bound and low computational complexity. In this chapter, we consider the more challenging case of multi-hop CR networks, where distributed algorithms are highly appealing. In [92], Ding and Xiao investigated the problem of enabling multisource video on-demand applications in multi-interface cognitive wireless mesh networks. Both centralized and distributed algorithms are developed for joint multipath routing and spectrum allocation for video sessions, aiming to minimize its total bandwidth cost in the network.

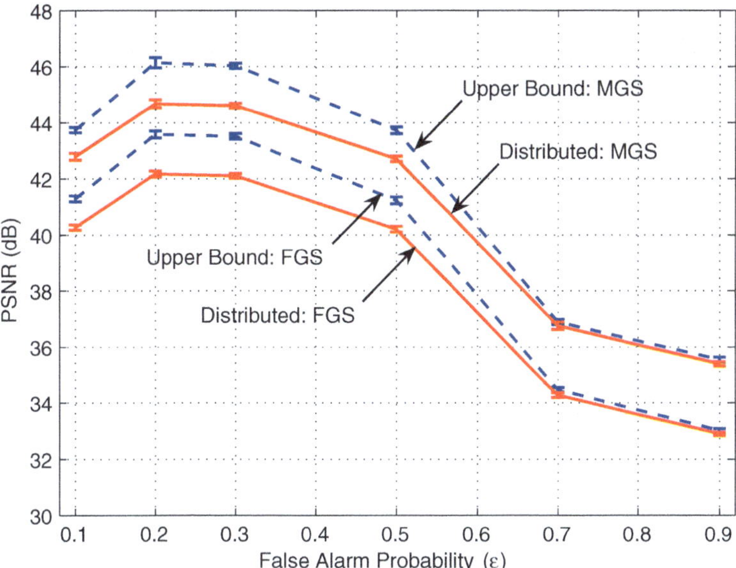

Fig. 5.9 Comparison of MPEG-4 FGS video with H.264/SVC MGS video under various false alarm probabilities ([4], ©2010 IEEE)

5.6 Conclusion

In this chapter, we studied the challenging problem of streaming multiple scalable videos in a multi-hop CR network. The problem formulation considered spectrum sensing and sensing errors, spectrum access and primary user protection, video quality and fairness, and channel/path selection for concurrent video sessions. We first solved the formulated MINLP problem using a sequential fixing scheme that produces lower and upper bounds on the achievable video quality. We then applied dual decomposition to derive a distributed algorithm, and analyzed its optimality and convergence performance. Our simulations validated the efficacy of the proposed scheme.

Chapter 6
Conclusion and Open Problems

6.1 Summary

Wireless video has been a challenging area with considerable research efforts. However, the problem of video communications over emerging CR networks has not been well studied, since the mainstream CR research has been focused on spectrum sensing and access. It was not clear whether video could be offered in such highly dynamic networks even a few years ago. There is a compelling need for innovative research in this area, given the Cisco prediction of exploding wireless video traffic increase in the next few years.

In this book, we investigated the problem of effective CR networking with application to multi-user video communications over four emerging CR networking paradigms, including infrastructure-based cellular CR networks, relay-assisted CR networks, femtocell CR networks, and multi-hop infrastructureless CR networks. This research involves network modeling, performance analysis, cross-layer design and optimization, algorithm development, and simulation validation. We aim to provide an in-depth treatment of the video over CR network problem with both theory and algorithm ingredients. The findings not only successfully demonstrated the feasibility of video CR networks, but also shed useful insights into developing practical CR video systems.

6.2 Open Problems and Future Work

Although considerable progresses are being made in the area of video over CR networks, there are many interesting open problems to be explored in this important problem area. Some such open problems are briefly described below, which are well worth being investigated in the future research.

First, in most of the prior work, it is assumed that the occupancy of each licensed channel evolves over time following a two-state discrete-time Markov process and the

S. Mao, *Video over Cognitive Radio Networks*, DOI: 10.1007/978-1-4614-4957-7_6,

primary user activities on different channels are independent. Although this assumption makes the problem manageable, it may not hold true in certain CR networks. The primary user transmission may be modeled as a more general process, while the primary transmissions on different channels may be statistically correlated. Thus, a more sophisticate spectrum sensing and access scheme is required to be integrated into the cross-layer optimization framework. The accuracy of the sensing process could be improved by exploiting the sensing results from adjacent channels and historic sensing results.

Similarly, another fundamental assumption made in this book is that the lowest video quality requirement for CR users can always be guaranteed. However, the network capacity that is available for CR users strongly depends on both primary user transmissions and randomly fading and shadowing channels. The CR network will eventually be congested and no video can be supported, when either there are too many CR video sessions or the primary users become highly active. Therefore, an admission control mechanism is required that can estimate the level of QoS that a new video session will have and whether there is enough bandwidth available to serve that session. A simple yet efficient admission control mechanism that considers both primary user activities and channel conditions is essential for QoS provisioning for video over CR networks.

Furthermore, we presented a theoretical framework for video streaming in CR networks and demonstrated the performance with extensive simulations. For future work, it would also be interesting to develop CR video streaming testbeds, such that the system performance can be demonstrated under a realistic wireless environment. This research can be focused on the combination of hardware components (e.g., the GNU Radio/Universal Software Radio Peripheral (USRP) platform) and software techniques (e.g., network optimization algorithms). Such a CR video testbed can not only validate the theoretical results, but also reveal new practical constraints that should be considered in the modeling and analysis, as well as identifying new research problems.

In addition, we focused on maximizing the PSNR of reconstructed video in the previous chapters, which is regarded as Quality of Service (QoS) provisioning. It is worth noting that QoS is from the network's perspective, which emphasizes the physical parameters such as delay, packet loss, Peak Signal-to-Noise Ratio (PSNR), etc. On the other hand, Quality of Experience (QoE) refers to the performance measures that can represent the end user's experience of viewing the reconstructed video, which are subjective measures [93]. In particular, different users may have different experiences with the same content and QoS level. It would be interesting to incorporate accurate yet manageable QoE models into the cross-layer optimization framework for QoE provisioning.

Last but not least, the case of coexistence of heterogeneous multimedia applications should be investigated. In this book, we focused on several challenging problems in CR networks using video as a reference application. In an operating CR network, there will be a plethora of different applications that generate different types of traffic flows, all sharing the extra bandwidth harvested by CRs. It is thus interesting to investigate how to provide QoS or QoE guarantees for heterogeneous traffic flows with different characteristics and different QoS/QoE requirements. This is a general problem for both wireline and wireless networks. The Internet adopts the Integrated Services(intserv) [94] and Differentiated Services (diffserv) [95] approaches to address this problem. We conjecture that a certain classification scheme should be adopted to intelligently identify and classify application traffic according to their characteristics and QoS/QoE requirements, and a resource allocation scheme will be used to differentiate the treatment of different classes of traffic flows.

These are interesting problems that are worth further investigation. We hope this monograph can serve as a starting point to trigger more interests in the community and attract more talented researchers to work in this interesting and important problem area. Such efforts will certainly help to meet the need for ubiquitous wireless video access, and generate considerable research advances and business opportunities.

Bibliography

1. D. Hu, S. Mao, Y. Hou, J. Reed, Fine grained scalability video multicast in cognitive radio networks. IEEE J. Sel. Areas Commun. **29**(3), 334–344 (2010)
2. D. Hu, S. Mao, Cooperative relay with interference alignment for video over cognitive radio networks, in *Proceedings of the IEEE INFOCOM'12*, Orlando, March 2012
3. D. Hu, S. Mao, On medium grain scalable video streaming over cognitive radio femtocell networks. IEEE J. Sel. Areas Commun. **30**(4) (2012)
4. D. Hu, S. Mao, Streaming scalable videos over multi-hop cognitive radio networks. IEEE Trans. Wireless Commun. **9**(11), 3501–3511 (2010)
5. Cisco, Cisco visual networking index: Global mobile data traffic forecast update, 2011–2016 (online). Available: http://www.cisco.com/en/US/solutions/collateral/ns341/ns525/ns537/ns705/ns827/white_paper_c11-520862.html. Feb 2011
6. S. Haykin, Cognitive radio: brain-empowered wireless communications. IEEE J. Sel. Areas Commun. **23**(2), 201–220 (2005)
7. J.M. Iii, Cognitive radio: An integrated agent architecture for software defined radio. Ph.D. dissertation, Royal Institute of Technology, Stockholm, Sweden, 2000
8. J. Reed, *Software Radio: A Modern Approach to Radio Engineering*, 1st edn. (Prentice Hall, Upper Saddle River, NJ, 2002)
9. FCC, Facilitating opportunities for flexible, efficient, and reliable spectrum use employing cognitive radio technologies. Notice of Proposed Rule Making and Order, FCC 03–322, Dec 2003
10. IEEE, Draft standard for wireless regional area networks part 22: cognitive wireless RAN medium access control (MAC) and physical layer (PHY) specifications: policies and procedures for operation in the TV bands. IEEE P802.22 Draft Standard (D0.3), May 2007
11. FCC, Report of the spectrum efficiency working group (online). Available: http://www.fcc.gov/sptf/reports.html. Nov 2002
12. Q. Zhao, B. Sadler, A survey of dynamic spectrum access. IEEE Signal Process. Mag. **24**(3), 79–89 (2007)
13. I. Akyildiz, W. Lee, M. Vuran, S. Mohanty, Next generation/dynamic spectrum access/cognitive radio wireless networks: a survey. Elsevier Comput. Netw. **50**(13), 2127–2159 (2006)
14. M. Dohler, R. Heath, A. Lozano, C. Papadias, R. Valenzuela, Is the PHY layer dead? IEEE Commun. Mag. **49**(4), 159–165 (2011)
15. H. Radha, M. van der Schaar, Y. Chen, The MPEG-4 fine-grained scalable video coding method for multimedia streaming over IP. IEEE Trans. Multimedia **3**(1), 53–68 (2001)

S. Mao, *Video over Cognitive Radio Networks*, DOI: 10.1007/978-1-4614-4957-7,

16. S. Mao, S. Lin, S.S. Panwar, Y. Wang, E. Celebi, Video transport over ad hoc networks: multistream coding with multipath transport. IEEE J. Sel. Areas Commun. **21**(10), 1721–1737 (2003)
17. Y. Hou, Y. Shi, H. Sherali, Spectrum sharing for multi-hop networking with cognitive radios. IEEE J. Sel. Areas Commun. **26**(1), 146–155 (2008)
18. Q. Zhao, S. Geirhofer, L. Tong, B. Sadler, Opportunistic spectrum access via periodic channel sensing. IEEE Trans. Signal Process. **36**(2), 785–796 (2008)
19. S. Geirhofer, L. Tong, B. Sadler, Cognitive medium access: constraining interference based on experimental models. IEEE J. Sel. Areas Commun. **26**(1), 95–105 (2008)
20. J. Nonnenmacher, E. Biersack, Optimal multicast feedback, in *Proceedings of the IEEE INFOCOM'98*, San Francisco, CA, March–April 1998, pp. 964–971
21. U. Berthold, F. Jondral, Guidelines for designing OFDM overlay systems, in *Proceedings of the IEEE DySPAN'05*, Baltimore, MD, Nov 2005, pp. 626–629
22. H. Tang, Some physical layer issues of wide-band cognitive radio systems, in *Proceedings of the IEEE DySPAN'05*, Baltimore, MD, Nov 2005, pp. 151–159
23. Y. Chen, Q. Zhao, A. Swami, Joint design and separation principle for opportunistic spectrum access in the presence of sensing errors. IEEE Trans. Inf. Theory **54**(5), 2053–2071 (2008)
24. S. Deb, S. Jaiswal, K. Nagaraj, Real-time video multicast in WiMAX networks, in *Proceedings of the IEEE INFOCOM'08*, Phoenix, AZ, April 2008, pp. 1579–1587
25. F. Kelly, A. Maulloo, D. Tan, Rate control in communication networks: shadow prices, proportional fairness and stability. J. Operat. Res. Soc. **49**(3), 237–252 (1998)
26. Y. Hou, Y. Shi, H. Sherali, Optimal base station selection for anycast routing in wireless sensor networks. IEEE Trans. Veh. Technol. **55**(3), 813–821 (2006)
27. S. Kompella, S. Mao, Y.T. Hou, H.D. Sherali, On path selection and rate allocation for video in wireless mesh networks. IEEE ACM Trans. Netw. **17**(1), 212–224 (2009)
28. L.-C. Wang, A. Chen, D. Wei, A cognitive MAC protocol for QoS provisioning in overlaying ad hoc networks, in *Proceedings of the IEEE CCNC'07*, Las Vegas, NV, Jan 2007, pp. 1139–1143
29. S. Srinivasa, S. Jafar, The throughput potential of cognitive radio: a theoretical perspective. IEEE Commun. Mag. **45**(5), 73–79 (2007)
30. T. Weingart, D. Sicker, D. Grunwald, A statistical method for reconfiguration of cognitive radios. IEEE Wireless Commun. Mag. **14**(4), 34–40 (2007)
31. H. Su, X. Zhang, Cross-layer based opportunistic MAC protocols for QoS provisionings over cognitive radio wireless networks. IEEE J. Sel. Areas Commun. **26**(1), 118–129 (2008)
32. S. Jafar, S. Srinivasa, Capacity limits of cognitive radio with distributed and dynamic spectral activity. IEEE J. Sel. Areas Commun. **25**(3), 529–537 (2007)
33. C.-T. Chou, S.S. N, H. Kim, K. Shin, What and how much to gain by spectral agility. IEEE J. Sel. Areas Commun. **25**(3), 576–588 (2007)
34. A. Fattahi, F. Fu, M. van der Schaar, F. Paganni, Mechanism-based resource allocation for multimedia transmission over spectrum agile wireless networks. IEEE J. Sel. Areas Commun. **25**(3), 601–612 (2007)
35. S. Mao, X. Cheng, Y. Hou, H. Sherali, Multiple description video multicast in wireless ad hoc networks. ACM Kluwer Mobile Netw. Appl. J. **11**(1), 63–73 (2006)
36. W. Wei, A. Zakhor, Multiple tree video multicast over wireless ad hoc networks. IEEE Trans. Circuits Syst. Video Technol. **17**(1), 2–15 (2007)
37. M. van der Schaar, S. Krishnamachari, S. Choi, X. Xu, Adaptive cross-layer protection strategies for robust scalable video transmission over 802.11 WLANs. IEEE J. Sel. Areas Commun. **21**(10), 1752–1763 (2003)
38. Y. Zhao, S. Mao, J. Neel, J.H. Reed, Performance evaluation of cognitive radios: metrics, utility functions, and methodologies. Proc. IEEE Special Issue Cognit. Radio **97**(4), 642–659 (2009)
39. A. Sendonaris, E. Erkip, B. Aazhang, User cooperation diversity-part I: system description. IEEE Trans. Commun. **51**(11), 1927–1938 (2003)
40. N. Laneman, D. Tse, G. Wornell, Cooperative diversity in wireless networks: efficient protocols and outage behavior. IEEE Trans. Inf. Theory **50**(11), 3062–3080 (2004)

41. Q. Zhang, J. Jia, J. Zhang, Cooperative relay to improve diversity in cognitive radio networks. IEEE Commun. Mag. **47**(2), 111–117 (2009)
42. J. Jia, J. Zhang, Q. Zhang, Cooperative relay for cognitive radio networks, in *Proceedings of the IEEE INFOCOM'09*, April 2009, pp. 2304–2312
43. D. Hu, S. Mao, On cooperative relay in cognitive radio networks, EAI Endorsed Transactions on Mobile Communications and Applications.
44. Z. Guan, L. Ding, T. Melodia, D. Yuan, On the effect of cooperative relaying on the performance of video streaming applications in cognitive radio networks, in *Proceedings of the IEEE ICC 2011*, Kyoto, Japan, June 2011, pp. 1–6
45. H.-L. Lee, S. Han, Y.-J. Oh, S.-T. Hwang, H. Kim, Multiple-input multiple-output communication system. U.S. Patent 2008/0 107 202 A1, May 2008
46. Q. Spencer, A.L. Swindlehurst, M. Haardt, Zero-forcing methods for downlink spatial multiplexing in multiuser MIMO channels. IEEE Trans. Sig. Process. **52**(2), 461–471 (2004)
47. D. Tse, P. Viswanath, *Fundamentals of Wireless Communication* (Cambridge University Press, Cambridge, UK, 2005)
48. V. Cadambe, S.A. Jafar, Interference alignment and the degrees of freedom for the *k* user ntererence channel. IEEE Trans. Inf. Theory **54**(8), 3425–3441 (2008)
49. L. E. Li, R. Alimi, D. Shen, H. Viswanathan, Y. R. Yang, A general algorithm for interference alignment and cancellation in wireless networks, in *Proceedings of the IEEE INFOCOM 2010*, San Diego, CA, March 2010
50. S. Katti, S. Gollakota, and D. Katabi, Embracing wireless interference: analog network coding, in *Proceedings of the ACM SIGCOMM'07*, Kyoto, Honshu, Aug 2007, pp. 397–408
51. S. Gollakota, S. David, D. Katabi, Interference alignment and cancellation, in *Proceedings of the ACM SIGCOMM'09*, Barcelona, Spain, Aug 2009, pp. 159–170
52. G. Strang, *Introduction to Linear Algebra*, 4th edn. (Wellesley Cambridge Press, Wellesley, MA, 2009)
53. D.P. Bertsekas, *Nonlinear Programming*, 2nd edn. (Athena Scientific, Nashua, NH, 1999)
54. H. Mahmoud, T. Yücek, H. Arslan, OFDM for cognitive radio: merits and challenges. IEEE Wireless Commun. **16**(2), 6–14 (2009)
55. C. Corderio, K. Challapali, D. Birru, S. Shankar, IEEE 802.22: an introduction to the first wireless standard based on cognitive radios. J. Commun. **1**(1), 38–47 (2006)
56. I.K. Son, S. Mao, Design and optimization of a tiered wireless access network, in *Proceedings of the IEEE INFOCOM*, San Diego, CA, March 2010, pp. 1–9
57. M. Wien, H. Schwarz, T. Oelbaum, Performance analysis of SVC. IEEE Trans. Circuits Syst. Video Technol. **17**(9), 1194–1203 (2007)
58. H.-P. Shiang, M. van der Schaar, Dynamic channel selection for multi-user video streaming over cognitive radio networks, in *Proceeings of the IEEE ICIP'08*, San Diego, CA, Oct 2008, pp. 2316–2319
59. L. Ding, S. Pudlewski, T. Melodia, S. Batalama, J. Matyjas, M. Medley, Distributed spectrum sharing for video streaming in cognitive radio ad hoc networks, in *International Workshop on Cross-layer Design in Wireless Mobile Ad Hoc Networks*, Niagara Falls, Canada, 2009
60. H. Luo, S. Ci, D. Wu, A cross-layer design for the performance improvement of real-time video transmission of secondary users over cognitive radio networks. IEEE Trans. Circuits Syst. Video Technol. **21**(8), 1040–1048 (2011)
61. V. Chandrasekhar, J.G. Andrews, A. Gatherer, Femtocell networks: a survey. IEEE Commun. Mag. **46**(9), 59–67 (2008)
62. R. Kim, J.S. Kwak, K. Etemad, WiMAX femtocell: requirements, challenges, and solutions. IEEE Commun. Mag. **47**(9), 84–91 (2009)
63. T. Rappaport, *Wireless Communications: Principles and Practice*, 2nd edn. (Prentice Hall PTR, Indianapolis, IN, 2001)
64. Q. Zhang, S. Kassam, Finite-state markov model for rayleigh fading channels. IEEE Trans. Commun. **47**(11), 1688–1692 (1999)
65. Z. Wang, L. Lu, A.C. Bovik, Video quality assessment using structural distortion measurement. Signal Process Image Commun. **2**, 121–132 (2004)

66. A. Hsu, D. Wei, C. Kuo, A cognitive MAC protocol using statistical channel allocation for wireless ad-hoc networks, in *Proceedings of the IEEE WCNC'07*, Hong Kong, P.R. China, March 2007, pp. 105–110
67. I. Guvenc, S. Saunders, O. Oyman, H. Claussen, A. Gatherer, Femtocell networks. EURASIP J. Wireless Commun. Netw. article ID 367878, 2 pages, doi:10.1155/2010/367878
68. V. Chandrasekhar, J.G. Andrews, T. Muharemovic, Z. Shen, A. Gatherer, Power control in two-tier femtocell networks. IEEE Trans. Wireless Commun. **8**(8), 4316–4328 (2009)
69. H.-C. Lee, D.-C. Oh, Y.-H. Lee, Mitigation of inter-femtocell interference with adaptive fractional frequency reuse, in *Proceedings of the IEEE ICC'10*, Cape Town, South Africa, May 2010, pp. 1–5
70. T. Alade, H. Zhu, J. Wang, Uplink co-channel interference analysis and cancellation in femtocell based distributed antenna system, in *Proceedings of the IEEE ICC'10*, Cape Town, South Africa, May 2010, pp. 1–5
71. J. O'Carroll, L.D. Holger Claussen, Partial GSM spectrum reuse for femtocells, in *Proceedings of the IEEE PIMRC'09*, Tokyo, Japan, Sept 2009, pp. 2111–2116
72. A. Galindo-Serrano, L. Giupponi, M. Dohler, Cognition and docition in OFDMA-based femtocell networks, in *Proceedings of the IEEE GLOBECOM'10*, Miami, FL, Dec 2010
73. F.-S. Chu, K.-C. Chen, Mitigation of macro-femto co-channel interference by spatial channel separation, in *Proceedings of the IEEE VTC-Spring'11*, Budapest, Hungary, May 2011, pp. 1–5
74. S. Rangan, Femto-macro cellular interference control with subband scheduling and interference cancelation, in *Proceedings of the IEEE GLOBECOM'10 Workshops*, Miami, FL, Dec 2010, pp. 695–700
75. Z. Bharucha, H. Haas, G. Auer, I. Cosovic, Femto-cell resource partitioning, in *Proceedings of the IEEE GLOBECOM Workshops*, Honolulu, HI, 2009
76. R. Madan, A. Sampath, A. Khandekar, J. Borran, N. Bhushan, Distributed interference management and scheduling in LTE-A femto networks, in *Proceedings of the IEEE GLOBECOM'10*, Miami, FL, Dec 2010, pp. 1–5
77. S.-M. Cheng, S.-Y. Lien, F.-S. Chu, K.-C. Chen, On exploiting cognitive radio to mitigate interference in macro/femto heterogeneous networks. IEEE Wireless Commun. Mag. **18**(3), 40–47 (2011)
78. S. Kaimaletu, R. Krishnan, S. Kalyani, N. Akhtar, B. Ramamurthi, Cognitive interference management in heterogeneous femto-macro cell networks, in *Proceedings of the IEEE ICC'11*, Kyoto, Honshu, June 2011, pp. 1–6
79. D. Hu, S. Mao, Multicast in femtocell networks: a successive interference cancellation approach, in *Proceedings of the IEEE GLOBECOM'11*, Huston, TX, Dec 2011, pp. 1–6
80. S. Ali, F. Yu, Cross-layer QoS provisioning for multimedia transmissions in cognitive radio networks, in *Proceedings of the IEEE WCNC'09*, Budapest, Hungary, April 2009, pp. 1–5
81. L. Ding, T. Melodia, S. Batalama, J. Matyjas, Distributed routing, relay selection, and spectrum allocation in cognitive and cooperative ad hoc networks, in *Proceedings of the IEEE SECON'10*, Boston, MA 2010
82. R. Kesavan, D.K. Panda, Efficient multicast on irregular switch-based cut-through networks with up-down routing. IEEE Trans. Parallel Distrib. Syst. **12**(8), 808–828 (2001)
83. R. Ramanathan, Challenges: a radically new architecture for next generation mobile ad-hoc networks, in *Proceedings of the ACM MobiCom'05*, Cologne, Germany, Sept 2005, pp. 132–139
84. J. Jia, Q. Zhang, X. Shen, HC-MAC: a hardware-constrained cognitive MAC for efficient spectrum management. IEEE J. Sel. Areas Commun. **26**(1), 106–117 (2008)
85. A. Motamedi, A. Bahai, MAC protocol design for spectrum-agile wireless networks: Stochastic control approach, in *Proceedings of the IEEE DySPAN'07*, Dublin, Ireland, April 2007, pp. 448–451
86. D. Hu, S. Mao, Design and analysis of a sensing error-aware MAC protocol for cognitive radio networks, in *Proceedings of the IEEE GLOBECOM'09*, Honolulu, HI, Nov–Dec 2009

87. R. Urgaonkar, M.J. Neely, Opportunistic scheduling with reliability guarantees in cognitive radio networks. IEEE Trans. Mobile Comput. **8**(6), 766–777 (2009)
88. T. Shu, M. Krunz, Throughput-efficient sequential channel sensing and probing in cognitive radio networks under sensing errors, in *Proceedings of the ACM MobiCom'09*, Beijing, China, Sept 2009, pp. 37–48
89. Y. Hou, Y. Shi, H. Sherali, Optimal spectrum sharing for multi-hop software defined radio networks, in *Proceedings of the IEEE INFOCOM'07*, Anchorage, AK, April 2007, pp. 1–9
90. Z. Feng, Y. Yang, Joint transport, routing and spectrum sharing optimization for wireless networks with frequency-agile radios, in *Proceedings of the IEEE INFOCOM'09*, Rio de Janeiro, Brazil, April 2009, pp. 1665–1673
91. D. Palomar, M. Chiang, A tutorial on decomposition methods for network utility maximization. IEEE J. Sel. Areas Commun. **24**(8), 1439–1451 (2006)
92. Y. Ding, L. Xiao, Video on-demand streaming in cognitive wireless mesh networks. IEEE Trans. Mobile Comput. **12**(3), 412–423 (2013)
93. A. Khan, L. Sun, E. Ifeachor, QoE prediction model and its application in video quality adaptation over UMTS networks. IEEE Trans. Multimedia **14**(2), 431–442 (2012)
94. B. Davie, C. Iturralde, D. Oran, S. Casner, J. Wroclawski, Integrated services in the presence of compressible flows. IETF RFC 3006 (online). Available: http://datatracker.ietf.org/doc/rfc3006/. Nov 2000
95. K. Chan, R. Sahita, S. Hahn, K. McCloghrie, Differentiated services quality of service policy information base. IETF RFC 3317 (online). Available: http://datatracker.ietf.org/doc/rfc3317/. March 2003

About the Author

Shiwen Mao received his Ph.D. in Electrical and Computer Engineering from Polytechnic University, Brooklyn, NY, USA (now Polytechnic Institute of New York University) in 2004. He was a research staff member with IBM China Research Lab from 1997 to 1998. He was a Postdoctoral Research Fellow/ Research Scientist in the Bradley Department of Electrical and Computer Engineering at Virginia Tech, Blacksburg, VA, USA from 2003 to 2006. Currently, he is the McWane Associate Professor in the Department of Electrical and Computer Engineering, Auburn University, Auburn, AL, USA.

His research interests include cross-layer optimization of wireless networks and multimedia communications, with current focus on cognitive radio, small cells, 60 GHz mmWave networks, Free Space Optical networks, and smart grid. He is on the Editorial Board of IEEE Transactions on Wireless Communications, IEEE Internet of Things Journal, IEEE Communications Surveys and Tutorials, Elsevier Ad Hoc Networks Journal, Wiley International Journal of Communication Systems, and ICST Transactions on Mobile Communications and Applications. He serves as the Director of E-Letter of Multimedia Communications Technical Committee (MMTC), IEEE Communications Society for 2012–2014. He serves as Technical Program Vice Chair for Information Systems (EDAS) of IEEE INFOCOM 2015, steering committee member of IEEE ICME and International Conference on Ad Hoc Networks, symposium co-chairs for many conferences, including IEEE ICC, IEEE GLOBECOM, ICCCN, IEEE ICIT-SSST, among others, and plays various roles in the organizing committees of many conferences.

S. Mao, *Video over Cognitive Radio Networks*, DOI: 10.1007/978-1-4614-4957-7,

Dr. Mao is a coauthor of *TCP/IP Essentials: A Lab-Based Approach* (Cambridge University Press, 2004) and co-editor of several books. He is a co-recipient of The IEEE ICC 2013 Best Paper Award. He received the 2013 IEEE ComSoc MMTC Outstanding Leadership Award and was named the 2012 Exemplary Editor of IEEE Communications Surveys and Tutorials. He was awarded the McWane Endowed Professorship in the Samuel Ginn College of Engineering for the Department of Electrical and Computer Engineering, Auburn University in August 2012. He received the US National Science Foundation Faculty Early Career Development Award (CAREER) in 2010. He is a co-recipient of The 2004 IEEE Communications Society Leonard G. Abraham Prize in the Field of Communications Systems. His paper was one of the runners-up for the 2013 Fabio Neri Best Paper Award, and he is a co-recipient of the Best Paper Runner-up Award at QShine 2008. He also received Auburn Alumni Council Research Awards for Excellence-Junior Award and two Auburn Author Awards in 2011. Dr. Mao holds one US patent. He is a member of Tau Beta Pi and Eta Kappa Nu and a senior member of the IEEE.

Index

S. Mao, *Video over Cognitive Radio Networks*, DOI: 10.1007/978-1-4614-4957-7,

The manufacturer's authorised representative in the EU is Springer Nature Customer Service Centre GmbH, Europaplatz 3, 69115 Heidelberg, Germany. If you have any concerns regarding our products, please contact ProductSafety@springernature.com

Printed and bound by CPI Group (UK) Ltd, Croydon, CR0 4YY
15/07/2026
02167645-0004